Concept Generation for Design Creativity

T0137954

Toshiharu Taura · Yukari Nagai

Concept Generation
for Design Creativity

A Systematized Theory and Methodology

 Springer

Toshiharu Taura
Department of Mechanical Engineering
Kobe University
Rokkodai-cho
Kobe 657-8501
Japan

Yukari Nagai
School of Knowledge Science
Japan Advanced Institute of Science and
  Technology
Asahidai 1-1
Nomi, Ishikawa 923-1292
Japan

ISBN 978-1-4471-5864-6   ISBN 978-1-4471-4081-8 (eBook)
DOI 10.1007/978-1-4471-4081-8
Springer London Heidelberg New York Dordrecht

Printed on acid-free paper

Springer is part of Springer Science+Business Media (www.springer.com)

Preface

This book discusses the very early stage of design. In order to help you understand what this book is about, we would like to explain the context of this book.

We will begin with a brief description of the authors, Toshiharu Taura and Yukari Nagai. When Toshiharu Taura was younger, he worked as a mechanical engineer for nearly 10 years. Back then, he believed that he experienced and captured the essence of 'design' in his mind. Since then, he has attempted to describe this essence using a more scientific approach—as a researcher—for nearly 20 years. Yukari Nagai studied 'art and design' at an art university, and she has been pursuing its 'heart' from both practical and theoretical perspectives, with an interest in creativity.

On the basis of our experience, we have been attempting to find an answer to one question—'what is design?'—for the last 10 years. It has been a long and difficult search. Nearly everyday was like this: sometimes, we felt as if we had caught hold of its (design's) tail, but it slithered away from us; at other times, when we felt we had caught it again, it turned out that we had caught merely its shadow.

Now, we hope that we have found an answer to this question, albeit concerning a very small part, and have, therefore, decided to concentrate all our efforts on writing a book entitled *Concept Generation for Design Creativity*.

The contents of this book are systematically laid out from theory to methodology and applications such that when a reader opens the book and begins reading it, through the flow of the book, he or she may understand the progression of our research.

Ten years ago, we would never have believed that our efforts would be systematized in the form of book chapters. In fact, the research in Chaps. 2, 3, and 4 had been conducted later than that in the other chapters. Incidentally, the first chapters were written last. Indeed, we are surprised to see this book compiled so systematically; all our prior efforts, which were attempted independently, were found to be related to each other. We have never 'designed' (i.e. in a well-planned, systematized manner) our research, but merely pursued the essence of design on the basis of our *inner sense*, which we believe is functioning to create the context of this book.

We would like to focus on the fact that our undesigned efforts have resulted in a book which looks designed. In other words, there may be another type of design which is not planned in a systematized manner; systematization is not the cause of design but a result of it. We would like to extend this message to you as our simple claim, which underlies the context of this book.

We hope this book will be read as an introduction to advanced studies on *concept generation* in design. To aid the reader in understanding this book, we have provided precise explanations of our research methods and procedures.

The main contributors to this book have been the co-authors of our previous publications, which are cited in this book; they were the ones who actually conducted the experiments or simulations. Our previous publications have been introduced in Chap. 1, and we would like to extend our gratitude to our co-authors. In addition, we would like to express our sincere gratitude to Dr. Georgi V. Georgiev and Dr. Kaori Yamada for their devoted contributions towards the editing of this book.

November 2011   Toshiharu Taura
                                                                                         Yukari Nagai

Contents

Chapter 1
Introduction

Abstract The aim, scope, and outline of this book are introduced in this chapter. The aim of this book is to develop a systematized theory and methodology on the thinking process at the very early stage of design in an interdisciplinary manner. We approach this aim from three perspectives: by focusing on the cognitive aspect, developing new research methods, and practising *concept generation*. In this chapter, first, the three perspectives are introduced. Next, the outline of the discussion in this book is presented. Finally, the organization of this book is explained and the authors' previous publications, which are related to this book, are listed.

1.1 Aim and Scope of This Book

This book attempts to develop a systematized theory and methodology on the thinking process at the 'very early stage of design' in an interdisciplinary manner, in order to answer the following questions: 'From where and how is a new concept generated?' and 'What enables a designer to promote the process to generate a new concept?' Here, 'very' is used to exaggerate the beginning of design, which includes the time just prior to or the precise beginning of the so-called conceptual design. At such a very early stage, a critical process seems to be shared among various design domains, such as engineering design and industrial design, whereas design research in existing academic fields, as well as design practice in actual business, is divided into each domain independently. Furthermore, this kind of stage is assumed to be located between the explicit cognition and implicit cognition, and thus, it is very difficult to capture its essence. In order to discuss this stage, various approaches from different academic disciplines, such as cognitive science, computer science, and mathematical consideration, should be gathered. Given that our attempt targets the process common to multiple design domains and approaches it from various academic disciplines, we use the term 'inter-disciplinarity'.

Furthermore, we use the term *concept* in this book. The reason is to capture the very early stage of design in a general manner by focusing on the image generated in a designer's mind (detailed definition in Chap. 2). In this book, the 'very early stage of design', that is, the stage just prior to or at the precise beginning of the so-called conceptual design phase, is called the process of **concept generation**.

Although we would like to capture the essence of the very early stage of design by using the notion of concept in an interdisciplinary manner, we cannot discuss this issue without setting a scope. The scope of this book is fixed as follows. First, we limit the scope of the concepts discussed in this book to those concepts related to 'things', and other concepts such as 'time' or 'space' are treated as beyond the scope. Here, 'things' involve nonphysical objects, such as software, pictures, and music, and physical objects—those which exist in a designer's mind as well as in the real world. Second, we assume that a new concept is not generated from nothing. This declaration involves two assumptions: that the basis of the *concept generation* exists, and a new concept is generated by referring to some existing concepts which lie either in the real world or in a designer's mind. However, we do not deny that a new concept might be generated suddenly in a designer's mind with no foretokening or basis, and we do not discuss the latter type of *concept generation* in this book owing to the difficulty in understanding these phenomena.

1.2 Outline of This Book

In this book, we approach the essence of *concept generation* from three per-spectives: by focusing on the cognitive aspect, developing new research methods, and practising *concept generation*.

1.2.1 Cognitive Approaches to Concept Generation

In order to capture the essence of *concept generation*, we will utilize the following approaches from the cognitive aspects: examine previous studies, create a model of *concept generation*, compare it with other processes which are different from the *concept generation*, and experimentally analyse the model.

The approach by examining previous studies is addressed in Chap. 2. First, the previous research on *concept generation* is overviewed by focusing on interdisci-plinarity and creativity. Next, *concept generation* is classified into two phases—the *problem–driven phase* and *inner sense–driven phase*—according to the following two factors: the basis of *concept generation* and ability which enables the *concept generation* to proceed. In addition, we define *concept generation* and creativity as follows: *concept generation* is the process of *composing* a desirable concept towards the *future*, and *design creativity* is the degree to which an *ideal* is conceptualized.

The approach which utilizes the modelling of *concept generation* is addressed in Chaps. 4 and 5. In Chap. 4, a systematized theory of *concept generation* is developed by comparing three existing theories—theories of metaphor and of abduction as well as General Design Theory (GDT) [13] from the viewpoint of similarities and dissimilarities. The theory classifies the *concept generation* into two types: *first-order concept generation* based on the 'similarity-recognition process' and *high-order concept generation* based on the 'dissimilarity-recognition process'. In Chap. 5, more specific methods of *concept synthesis* for *concept generation* are systematized, as follows: *property mapping, concept blending*, and *concept integration in thematic relation*. So far, these theories and methods have been discussed independently, and the relations among them have not been clarified. In this book, they are systematized by relating them to each other and structuring them in a unified manner.

Furthermore, the essential nature of *concept generation* is discussed in Chap. 3 in terms of 'competence'. The competence is categorized into three types: competence to inspire the motivation from inside a thought space, competence to abstract the concepts, and competence to control the *back-and-forth issue*.

The approach which utilizes a comparison with another process is addressed in Chap. 5. In this chapter, the experiment to compare *concept generation* with the process of linguistic interpretation is discussed in order to determine the characteristics which feature in the *concept generation*. In addition, we examine whether the recognition type is manifested during *concept generation* or derived from a trait accumulated in subjects.

The approach of experimental analysis is addressed in Chaps. 6 and 8 in addition to Chap. 5. In Chap. 6, the expansion of a thought space during creative *concept synthesis* is investigated. In Chap. 8, the conditions for successful *concept synthesis*, with a focus on the distance between the *base concepts*—which are used to generate a new concept—and the associativeness of the *base concepts*, are examined.

1.2.2 New Research Methods for Concept Generation

It is difficult to observe the thinking process of *concept generation* objectively, whether externally or internally, since a designer is assumed to be deeply engaged in his or her work during the creative process. Accordingly, developing a new research method and verifying its effectiveness should also be investigated. In this book, two methods are proposed. The first one is the *extended protocol analysis method*, which is introduced in Chap. 6. An experiment using this method is conducted in order to obtain the in-depth data pertaining to the expressed design activity. The second method is a computer simulation of the *concept synthesis*, which is introduced in Chap. 7. This simulation employs a research framework called *constructive simulation*, in which the virtual process is developed on a semantic network by tracing the relationships between its governing concepts.

1.2.3 Practice of Concept Generation

We also seek the essence of *concept generation* by applying the methods of
concept synthesis for more specific processes and by confirming its feasibility. In
this book, these methods are applied to three applications: shape design, motion
design, and function design. In Chap. 9, a method to synthesize the abstract shapes
is developed. In Chap. 10, a method to synthesize motions in order to generate a
creative and emotional motion is developed. Moreover, in Chap. 11, a method to
synthesize functions in order to support the design of a new function is developed.

1.2.4 Discussion

In Chap. 12, a complete discussion on the three themes is attempted in the following manner: 'What are the *inner criteria*?', 'What is an *ideal* in *concept
generation*?', and 'How should the very early stage of design be investigated?'

First, *inner criteria* in *inner sense* is discussed as follows. On the basis of the
results of the experiments, it is indicated that we have common *inner criteria* in
our minds which evaluate creativity and which give a reason to the existence of the
inner criteria. In addition, a similar kind of *inner criteria* is considered active in
the processes of *concept generation* and concept evaluation. Further, the paradox
that a variety of *design ideas* are generated from common *inner criteria* is
explained as follows: the *variety* in *design ideas* originates from how specific
concepts are used to implement the structure of the *concept generation* process,
whereas the creativity of the generated *design ideas* is evaluated on the basis of the
characteristics of the network structure without specific concepts.

Next, *ideal* is discussed as follows. On the basis of the considerations of *ideal* in
related fields, it is inferred that the notion of *ideal* may be different from that of
'existable' or 'achievable'; moreover, ideal design processes or ideal design
products should be those which need not necessarily exist.

Last, the methods of *concept generation* are discussed as follows: a break-
through from the stimulus–response framework needs to be developed and an ideal
design process, rather than inducing actual design processes conducted in the real
world, should be pursued.

1.2.5 Towards the Future

In Chap. 12, some perspectives towards the future of *concept generation* are
addressed.

First, we explain that the competence for *concept generation*, which is discussed in Chap. 3, is persistent in nature and is expected to contribute to an
extension which is beyond the scope of this book.

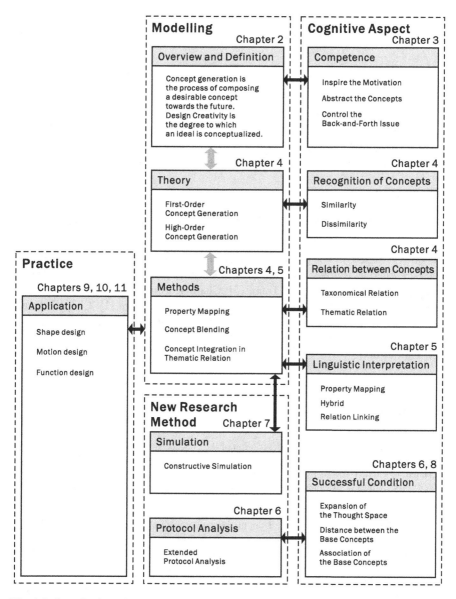

Fig. 1.1 Organization of this book

Next, we point out that the *inner sense–driven phase* is expected to contribute to the in-depth discussion on the next generation of design, in which we are released from a belief in efficiency, and other important meanings of design are anticipated to arise.

Finally, by considering *concept generation*, we may also attempt to answer the essential question: 'What is a human being?'

1.2.6 Organization of This Book

The organization of this book is presented in Fig. 1.1. As mentioned above, a theory and methodology of *concept generation* were discussed from three perspectives: cognitive aspects, new research methods, and practice.

1.2.7 Authors' Previous Publications Related to This Book

This book is based on our previous studies and refers to our previous publications which are cited as follows: Chaps. 2 and 3 are cited from the discourses of Taura and Nagai [6, 7] and Nagai and Taura [3]; Chap. 4, from Taura and Nagai [6, 8]; Chap. 5, from Nagai et al. [4]; Chap. 6, from Nagai and Taura [1]; Chap. 7, from Taura et al. [11]; Chap. 8, from Taura et al. [9] and Nagai and Taura [2]; Chap. 9, from Taura and Nakayama [10]; Chap. 10, from Yamada et al. [12]; and Chap. 11, from Park et al. [5].

References

1. Nagai Y, Taura T (2006) Formal description of the concept-synthesizing process for creative design. In: Gero JS (ed) Design computing and cognition '06. Springer, London
2. Nagai Y, Taura T (2006) Role of action concepts in the creative design process. In: Kim YS (ed) Proceedings of the international design research symposium. Seoul, November, pp 257–267
3. Nagai Y, Taura T (2010) Discussion on direction of design creativity research (part 2)—research issues and methodologies: from the viewpoint of deep feelings and desirable figure. In: Taura T, Nagai Y (eds) Design creativity 2010. Springer, London
4. Nagai Y, Taura T, Mukai F (2009) Concept blending and dissimilarity: factors for creative concept generation process. Des Stud 30:648–675
5. Park Y, Ohashi S, Yamamoto E, Taura T (2011) Design of functions by function blending. In: Culley SJ, Hicks BJ, McAloone TC, Howard TJ, Reich Y (eds) In: Proceedings of the 18th international conference on engineering design, impacting society through engineering design. Design theory and research methodology, vol. 2. Lyngby/Copenhagen, Denmark, 15–19 Aug, pp 61–72
6. Taura T, Nagai Y (2010) Concept generation and creativity in design. Cogn Stud 17:66–82 (in Japanese)
7. Taura T, Nagai Y (2010) Discussion on direction of design creativity research (part 1)—new definition of design and creativity: beyond the problem-solving paradigm. In: Taura T, Nagai Y (eds) Design creativity 2010. Springer, London

8. Taura T, Nagai Y (2013) A systematized theory of creative concept generation in design: first-order and high-order concept generation. Res Eng Des (under submission)
9. Taura T, Nagai Y, Tanaka S (2005) Design space blending—a key for creative design. In: Proceedings of the 15th international conference on engineering design. Melbourne, 15–18 Aug (CD-ROM)
10. Taura T, Nakayama Y (2004) A study on the representation method of abstract form features by evaluation functions—acquisition of the evaluation function for generating forms using evolutionary computation methods. Bull Jpn Soc Sci Des 51:35–44 (in Japanese)
11. Taura T, Yamamoto E, Fasiha MYN, Goka M, Mukai F, Nagai Y, Nakashima H (2012) Constructive simulation of creative concept generation process in design: a research method for difficult-to-observe design-thinking processes. J Eng Des 23:297–321. doi:10.1080/09544828.2011.637191
12. Yamada K, Taura T, Nagai Y (2010) Design of emotional and creative motion by focusing on rhythmic features. In: Taura T, Nagai Y (eds) Design creativity 2010. Springer, London
13. Yoshikawa H (1981) General design theory and a CAD System. In: Sata T, Warman EA (eds) Man-machine communication in CAD/CAM. North-Holland, Amsterdam

Chapter 2
Perspectives on Concept Generation and Design Creativity

Abstract In this chapter, we address the perspectives on design, *concept generation*, and creativity. First, we overview the previous research on *concept generation*, which is at the very early stage of design, by focusing on interdisciplinarity and creativity. Next, we classify *concept generation* into two phases—the *problem–driven phase* and *inner sense–driven phase*—according to the following two factors: the basis of *concept generation* and ability which enables the *concept generation* to proceed. Further, we define *concept generation* and creativity as follows: *concept generation* is the process of *composing* a desirable concept towards the *future*, and *design creativity* is the degree to which an *ideal* is conceptualized.

2.1 Very Early Stage of Design

In the field of design research, design is usually described as an activity to formulate a solution for a purpose [62]. Indeed, the process of design has been understood to be the process of rational problem solving by transforming existing situations into preferred ones [11, 41]. On the other hand, design has been viewed as a 'reflective practice' [58], and a number of studies have investigated design processes in order to identify the features of the thinking process [15, 20, 24, 48]. These studies have identified interesting issues in the design process such as rationality, expertise, and learning. In this book, we approach the essential phenomenon in the characteristics of creativity in design by focusing on the 'very' early stage of design which includes the time just prior to or the precise beginning of the so-called conceptual design.

The meanings of design typically involve two phases: the mental plan for something, followed by the creation of forms. The former phase is generally termed the conceptual design. In engineering design, conceptual design has been considered an 'early stage of design' in a systematic approach [49]. In this approach, schematic descriptions of a mechanism are important at the early stage to initialize the flow of the design process. Furthermore, methods for a systematic approach have been developed to determine the relations between a mechanism and functions [39, 45, 64]. In these processes, originality as well as rationality is required [2]. Creative techniques for systematic design such as checklist methods—for example, the Attribute Listing Method [12] based on the decomposition of the problem—as well as network modelling methods—for example, the Option Graph Method [44] based on the linkages of the elements in a flow diagram—were adopted in order to find new solutions.

Industrial design can be considered that which aims to create admired shapes and colours and which has been historically developed from art and craftwork. In industrial design, the early stage of design played the role of being an ideation process for forming whilst drawing [60, 65]. In the stage of conceptual design, a suitable goal of product design is found as a result of ideation, which will also answer to social requirements [66, 71]. Market research and surveys on product trends have been believed to be effective supplemental methods for the ideation of a product design [7, 10]. These investigations provide information on the gap between users' or customers' needs and the existing product in a market. Although the market research and surveys are directed towards finding the goal of product design, conceptualizing new ideas for products (novel products), which cannot be obtained through market research and surveys is thought to be another crucial approach; however, this process is still difficult and thought to stem from an inspiration.

Indeed, the critical process to produce a novel product at the very early stage of design is assumed to be shared between the multiple design domains (engineering design, industrial design, etc.) as an unsystematized phenomenon. As mentioned in Chap. 1, in this book, the very early stage of design, during which an initial idea or specification is generated, is called *concept generation*.

2.2 Interdisciplinary View of Research on Concept Generation

Some studies on modelling ideation processes corresponded to *concept generation*; for instance, those of engineering design and industrial design [5, 6, 13, 37, 38, 43, 51] as well as architectural design [1, 11]. The studies highlighted each aspect of *concept generation*.

To assist the *concept generation*, two types of methodological support techniques have been developed: the visual method and linguistic method. The visual method type is usually based on visual and spatial cognition using imagery resources or graphical media [17, 21, 31, 46, 67], including 3-dimensional design [16, 53] and virtual information [50]. The visual method type is thought to be effective in assisting a designer's (or design team's) image aspect of *concept generation* in the shape, interface, or usage scene of a product for industrial design, as well as in the mechanical aspect for engineering design. The linguistic method type is based on language and uses lexicon technology [9, 42, 56]; it is supposed to contribute more towards activating *concept generation* at the abstract level, such as the meanings or social values of a product. Both types are considered useful for accelerating or efficiently driving *concept generation*.

With regard to another methodology on *concept generation*, Brainstorming [47] is a popular method to facilitate the ideations quantitatively and is introduced to obtain the frequency of finding new solutions; however, the effectiveness of Brainstorming in engineering design is argued [36, 59, 69], and the qualitative limitations of Brainstorming have been addressed [73].

In addition, other general creativity support techniques are proposed to strengthen the originality in ideation process by providing a new method of seeing a situation. Synectics [29] is a famous example of an operational theory for the conscious use of the psychological mechanisms of creative activity, particularly with regard to the roles of metaphor in the creative process. Later, the roles of metaphor provided a computational cognitive model of analogy [8, 33, 35] such as Copycat [32, 34]. Empirically, analogy is viewed as a creative strategy for shifting the manner in which a situation is seen and is thought to be an effective stimulus for design [20, 28, 36]. For developing multiple viewpoints on seeing situations, flexibility is paid attention to and 'lateral thinking' [19] is introduced as an effective creative thinking technique which is related to a flexibility developing skill. This is further developed into the Six Thinking Hats [18] tool for human decision making. In order to obtain the meta-level skill of seeing a situation, not only human thinking but also the need for 'learning by doing' is highlighted from the viewpoint of practice. In the framework of learning in practice theory, the skill of reflection is introduced as an effective method to develop the creativity of 'practitioners' [57, 58].

To improve the ideation support methods mentioned above, it is necessary to identify the theory of *concept generation* by gauging our understanding of human cognitive features including inspiration. Indeed, the generative process lying behind mysterious phenomena and its ability at the very early stage of design should be an interdisciplinary (beyond the existing academic disciplines and across multiple design domains) research theme. In this book, we attempt to develop a systematized theory and methodology on *concept generation*, in particular, in an interdisciplinary manner.

2.3 Creativity in Design

In the field of design research, two kinds of creativity have been discussed. One is related to the process of design, while the other is related to the products that are the outcomes of the design process.

In the former, the emphasis is sometimes on rational decision making to find a design solution within the framework of problem solving. Alternatively, Cross [14] reports many cases of creative leaps that have been made during the design process, which may have been caused by the release from a mental fixation. The role of visual information is considered conducive to such releases from mental fixations. In fact, it has been supposed that experts have actual knowledge of how to break such fixations. Until now, analogical reasoning has been given the most attention because it is related to the breaking of fixations [4, 22, 26, 70]. Many studies have reported that metaphors and visual images are effective for analogical reasoning [27], and expert designers seem to understand the roles of metaphors and visual images. Referring to Arnheim's [3] theory, these studies have claimed that the capacity for visual thinking might be particularly expanded in the cognitive process of designers. Goldschmidt [27] identifies the effects of the ideas that occur in visual thinking on the abstraction level during the design process, which she relates to creativity, by carrying out experimental studies. These results were obtained through experimental observation of architectures' design protocols. Such an experimental observation of the design process has been called 'design protocol studies', and it provides information on the cognitive features of the creative design process [40].

On the other hand, the creativity of designed outcomes or the ideas governing them have usually been evaluated in terms of novelty and practicality—the two criteria offered by the study of Sternberg and Lubart [63]. They describe creativity as the ability to produce work with both novelty and appropriateness. Weisberg [72] points out the importance of 'values' for creativity evaluation. Vargas–Hernandez et al. [68] evaluate the sketches as the outcomes of the design process with two criteria: novelty and quality. Sarkar and Chakrabarti [55] propose the assessment methods of 'novelty and usefulness' in order to evaluate the values of designed outcomes. Gero [25] adds the notion of 'unexpectedness' to these criteria.

Furthermore, there are other standards with which to evaluate creativity. For instance, the value of the diversity of products or the speed at which goals are achieved is often used as a criterion for creativity evaluation (in the *Encyclopedia of Creativity* 1999) [54]. In addition, Ulrich and Eppinger [66] suggest that the actual marketing results imply the values of creativity in the real world. They also suggest that diversity of productions affect the power to create products in the next generation.

As mentioned above, many studies have discussed the creativity in design; however, further consideration is thought to be necessary to capture the essence of creativity at the very early stage of design, that is, in *concept generation*. In accordance with our attempt to develop an interdisciplinary theory and methodology for *concept generation*, we also aim to identify the creativity for *concept generation* in the following section (in this book, the creativity for *concept generation* is referred to as ***design creativity***).

2.4 Phases of Concept Generation

To specify what *concept generation* is, we classify the process of *concept generation* into two phases—the *problem–driven phase* and *inner sense–driven phase*—according to the following two factors: the basis of the *concept generation* and ability which enables the *concept generation* to proceed. These factors are related to the following questions, respectively: (1) From where is a new concept generated? and (2) What enables a designer to promote the process to generate a new concept?

2.4.1 Concept

In this book, **concept** is defined as that which refers to the figure of an object, along with other representations such as attributes or functions of the object, which existed, is existing, or might exist in the human mind as well as in the real world. This definition is in line with previous considerations in the field of design study (e.g. [11, 30, 52, 61]). Here, 'figure' implies the notion of an image as well as a physical shape, and 'object' involves not only a physical object but also a non-physical object: software, music, and so on. As mentioned in Chap. 1, we assume that a new concept is not generated from nothing. This declaration involves two assumptions: that the basis of generating a new concept exists, and a new concept is generated by referring to some existing concepts which lie either in the real world or in a designer's mind. However, we do not deny that a new concept might be generated suddenly in the designer's mind with no foretokening or basis, and we do not discuss this type of *concept generation* in this book owing to the difficulty in understanding these phenomena.

2.4.2 Problem–Driven Phase

We call a gap that exists between a goal of an object and its existing situation a **problem**, and define the **problem–driven phase** as the process of generating a new concept (solution) on the basis of the *problem*. In certain situations, there are obvious goals that need to be achieved, such as finding solutions for natural disasters. Similarly, in cases where we need to meet customers' explicit require-ments, it is also easy to set goals. In the *problem–driven phase*, the new concept (solution) can usually be obtained by analysing the *problem* (this point will be discussed in more detail in Chap. 4).

Through these definitions and considerations, we can deduce that the basis of the concept is a *problem* (the gap between a goal of an object and its existing situation) and that the ability which enables this phase to proceed is that of 'analysing'.

Next, we discuss the creativity of the *problem–driven phase*. That the solution can be obtained by analysing the gap indicates that the solution lies hidden in the gap. This consideration suggests that the creativity of the *problem–driven phase* is related to the process of discovering the hidden solution in the gap.

2.4.3 Inner Sense–Driven Phase

The **inner sense–driven phase** is defined as the process of generating a new concept on the basis of the *inner sense* for pursuing an *ideal*. Although the notion of the *inner sense* in design has been recognized in previous research (e.g. [30, 61]), further discussion on this aspect will aid in understanding and systematizing *concept generation*. Here, **inner sense** is that which involves *inner criteria* and 'intrinsic motivation' and can be the basis on which a new concept is generated by referring to existing concepts; *inner criteria* are that which is explicitly or implicitly underlying in the designer's mind and guides the process of *concept generation*; this issue will be discussed again in Chap. 12, and 'intrinsic motivation' will be addressed in Chap. 3.

An *ideal* is considered the direction pursued by the *inner sense*. In other words, in the *inner sense–driven phase*, the most important element could be the generation of a new concept for pursuing an *ideal*. From the viewpoint of engineering design, the development of the ideal functions of future artefacts is implied, while from the viewpoint of industrial design, the development of the ideal shapes or interfaces which evoke an ideal impression on a user's mind is implied.

In some cases, when an *ideal* is explicitly expressed, the *ideal* may become a 'goal' in the *problem–driven phase*. However, the notion of *ideal* in this phase is not to be approached by analysing the current state. If it can be easily obtained from an analysis of the current state, it should be categorized into the *problem–driven phase*. To approach an *ideal*, the ability of *composing*, which is the notion opposed to analysis, is considered inevitable. In the very early stage of design, one of the typical processes is the composition of elements, because the way in which we create products differs from the process of creation in the natural world. It is a well-known fact that such a composition of elements is observed in the human recognition process [23]. The ability of *composing* is also effective in the *problem–driven phase*; however, we believe this ability features in the *inner sense–driven phase*. As a notion opposed to 'analysis', 'synthesis' is also well-known. In this book, the term 'synthesis' is used later (in Chap. 5) in the more specific context.

Further, the creativity of the *inner sense–driven phase* is assumed to be related to the process of approaching an *ideal* through the composition of elements.

Table 2.1 summarizes the two phases of the *concept generation* described.

The relationship between the two phases is illustrated in Fig. 2.1.

In the actual design process, these two phases do not work independently; instead, they realize the design process complementarily. In the design process which is modelled in the framework of the so-called problem-solving process [49],

Table 2.1 The two phases of concept generation with two factors: the basis and ability of concept generation

Phase of concept generation	Basis	Ability
Problem–driven phase	Problem	Analysing
Inner sense–driven phase	Inner sense	Composing

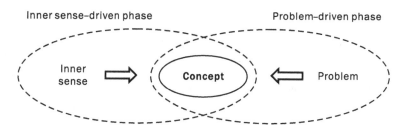

Fig. 2.1 The two phases in concept generation

a new concept through which the given goal is achieved is generated by analysing the gap between the goal and the current situation; moreover, it is achieved by creating a new solution or decomposing the *problem* on the basis of the designer's *inner sense*.

On the other hand, even if the design is captured as an activity of the artist, its very early stage can also be understood to comprise the two phases, where focus is placed on the *inner sense–driven phase*. In the very early stage of art-like design, a new concept is thought to be generated from the designer's *inner sense,* followed by fitting it into the situation by analysing the gap between the emergent vague concept and the current situation.

2.5   Definitions of Concept Generation and Design Creativity

On the basis of the above considerations, we define ***concept generation*** as *the process of composing a desirable concept towards the future.*

This definition aims at developing a framework in which the *concept generation* can be structured in an interdisciplinary manner, while focusing on our view of *concept generation*. In our previous studies, the *inner sense–driven phase* was found to be related to the essence of the very early stage of design (this point will be discussed in Chaps. 4–6), even though the relation is indirect. Accordingly, we focus on the *inner sense–driven phase*; however, this focus intends not to narrow the meaning of *concept generation*, but to clarify our stand to capture the essence of *concept generation*, which is the main theme of this book.

With regard to this definition, we will first explain what we mean by the phrase 'towards the *future*'. The notion of the *future* is thought to be, of course, extremely abstract, as follows. We can never draw an exact picture of the *future*. We can imagine what things may be like in the future, but it is impossible to visualize a precise notion of the *future* itself. We think this kind of highly abstract notion can only be represented in language. That is, the notion of *future* is considered to be recognized only by human beings through the use of language. On the other hand, *concept generation* is also a highly intellectual activity. Accordingly, these two notions of *future* and *concept generation* are believed to feature in human beings. Therefore, we would like to attempt to describe this part in order to identify the fact that *concept generation* characterizes human beings, whereas this characteristic is thought to be straightforward owing to the fact that every activity of living things is directed towards the future. In the context of *concept generation*, the *future* is considered to have two meanings. One meaning is a future that we can grasp inductively, such as a marketing forecast. The other meaning is a future that is recognized in the wish or desire for creation which is led by an *inner sense*. In our understanding of *concept generation*, we consider that the latter meaning is the more essential of the two.

Next, we explain what we mean by 'desirable concept'. It is this part which determines the object of *concept generation*. Here, the notion of 'desirable' could imply 'the pursuit of certain *ideals*', which will never be realized in the real world, as well as the 'implementation of a wish', which can be realized in the real world. Accordingly, it is suggested that there are two kinds of desirable concepts: an ideal shape or function in the *inner sense–driven phase* and obvious given goals in the *problem–driven phase*. As mentioned above, in our understanding of *concept generation*, we consider this former object to be more essential. In this case, we think a *sense of resonance* in the mind can be an *inner criterion* for an *ideality*. One important point regarding artefacts is the notion of 'naturalness'. We often assume that the process of making artefacts should come naturally to humans. However, there is no credible process that *resonates* with human beings, even though we create artefacts by copying them from the natural world. In contrast, there are some things that differ from what is found in the natural world but nevertheless, 'naturally' *resonate* with the human mind. Music is a good example. Music is composed of man-made sounds, most of which differ from the sounds which exist in the natural world, such as the sound of a breeze or a bird's song. Indeed, music *resonates* with the human mind.

Finally, we explain what we mean by '*composing*'; it is this part which determines the process of *concept generation*. For the composition, particularly in the composition of desirable concepts, finding a new concept which can be realized is thought to be insufficient. We assume it is also necessary to pursue the desirable concept towards the *future* which will never be realized, and there must be an intrinsic motivation in one's *inner sense* for this pursuit. In this book, we use the term *composing* in order to emphasize the meaning of embracing such motivations with one's *inner sense* to pursue an *ideal* in addition to the normal meaning mentioned above.

On the basis of the above considerations, we define ***design creativity*** in *concept generation* as <u>*the degree to which an ideal is conceptualized*</u>.

In this definition, novelty may be implemented as a by-product of *concept generation*, but not as a causal factor of creativity. Thus, if a new concept is pursued merely on account of its uniqueness, we say that this pursuit never approaches an *ideal*.

On the other hand, hereafter in this book, the unadorned term 'creativity' suggests the conventional meaning: originality and practicality.

With regard to the meaning of *ideal* and *inner sense* in *concept generation* and the relevance of this definition, we will discuss them again in Chap. 12.

References

1. Aliakseyeu D, Martens JB, Rauterberg M (2006) A computer support tool for the early stages of architectural design. Interact Comput 18:528–555. doi:10.1016/j.intcom.2005.11.010
2. Archer B (1965) Systematic method for designers. The Design Council, London
3. Arnheim R (1969) Visual thinking. University of California Press, Berkeley
4. Ball LJ, Christensen BT (2009) Analogical reasoning and mental simulation in design—two strategies linked to uncertainty resolution. Des Stud 30:169–186. doi:10.1016/j.destud.2008.12.005
5. Bilda Z, Demirkan H (2003) An insight on designers' sketching activities in traditional versus digital media. Des Stud 24:27–50. doi:10.1016/S0142-694X(02)00032-7
6. Bilda Z, Gero JS, Purcell T (2006) To sketch or not to sketch? that is the question. Des Stud 27:587–613. doi:10.1016/j.destud.2006.02.002
7. Bloch PH (1995) Seeking the ideal form: product design and consumer response. J Mark 59:16–29. doi:10.2307/1252116
8. Boden MA (1998) Creativity and artificial intelligence. Artif Intell 103:347–356. doi:10.1016/S0004-3702(98)00055-1
9. Chiu I, Shu LH (2007) Using language as related stimuli for concept generation. AI EDAM 21:103–121. doi:10.1017/S0890060407070175
10. Cooper RG, Kleinschmidt EJ (1988) Resource allocation in the new product process. Ind Mark Manage 17:249–262. doi:10.1016/0019-8501(88)90008-9
11. Coyne RD, Rosenman M, Radford D, Balachandran M, Gero JS (1990) Knowledge-based design systems. Addison-Wesley, Reading
12. Crawford RP (1964) The techniques of creative thinking—how to use your ideas to achieve success. Fraser publishing, Wells
13. Cross N (2001) Design cognition: results from protocol and other empirical studies of design activity. In: Eastman C, Newstetter W, McCracken M (eds) Design knowing and learning: cognition in design education. Elsevier, Amsterdam. doi:10.1016/B978-008043868-9/50005-X
14. Cross N (2006) Designerly ways of knowing. Birkhäuser, Basel
15. Cross N, Christiaans H, Dorst K (1996) Analysing design activity. Wiley, Chichester
16. Dagman A, Söderberg R, Lindkvist L (2007) Split-line design for given geometry and location schemes. J Eng Des 18:373–388. doi:10.1080/09544820601008969
17. Dahl DW, Chattopadhyay A, Gorn GJ (1999) The use of visual mental imagery in new product design. J Mark Res 36:18–28. doi:10.2307/3151912
18. de Bono E (1999) Six thinking hats. Back Bay Books by Little, Brown and Co, New York
19. de Bono E (2009) Lateral thinking: a textbook of creativity. Penguin, London

20. Dorst K, Cross N (2001) Creativity in the design process: co-evolution of problem–solution. Des Stud 22:425–437. doi:10.1016/S0142-694X(01)00009-6

21. Edmonds EA, Soufi B (1994) Perceptual interpretation and representation of emergent shapes. Preprints of workshop on reasoning with shapes in design. In: Proceedings of the 3rd international conference on artificial intelligence in design. University of Sydney and the Federal Institute of Technology, Switzerland, pp 39–45

22. Findler NV (1981) Analogical reasoning in design process. Des Stud 2:45–51. doi:10.1016/0142-694X(81)90029-6

23. Finke RA, Ward TB, Smith SM (1992) Creative cognition: theory, research, and applications. The MIT Press, Cambridge

24. Friedman K (2000) Design knowledge: context, content and continuity. In: Durling D, Friedman K (eds) Proceedings of the La Clusaz conference, foundations for the future, Doctoral education in design. Staffordshire University Press, Staffordshire, UK, pp 8–12 July

25. Gero JS (2007) on his talk 'Design creativity', ICED07. SIG Design Creativity Workshop

26. Goldschmidt G (1990) Linkography: assessing design productivity. In: Trappl R (ed) Cybernetics and systems 90. World Scientific, Singapore, pp 291–298

27. Goldschmidt G (1994) On visual design thinking: the vis kids of architecture. Des Stud 15:158–174. doi:10.1016/0142-694X(94)90022-1

28. Goldschmidt G (2001) Visual analogy—a strategy for design reasoning and learning. In: East-man C, Newstetter W, McCracken M (eds) Design knowing and learning: cognition in design education. Elsevier, Amsterdam. doi: 10.1016/B978-008043868-9/50009-7

29. Gordon WJJ (1961) Synectics. Harper & Row, New York

30. Hatchuel A, Weil B (2009) C-K design theory: an advanced formulation. Res Eng Des 19:181–192. doi:10.1007/s00163-008-0043-4

31. Herring SR, Chang CC, Krantzler J, Bailey BP (2009) Getting inspired! understanding how and why examples are used in creative design practice. Proceedings of the 27th international conference on human factors in computing systems. ACM Press, New York. doi: 10.1145/1518701.1518717

32. Hofstadter D (1993) How could a copycat ever be creative? AAAI Tech Rep SS 93:1–10

33. Hofstadter D (1995) Fluid concepts and creative analogies: computer models of the fundamental mechanisms of thought. Basic Books, New York

34. Hofstadter D (2001) Analogy as the core of cognition. In: Gentner D, Holyoak K, Kokinov B (eds) The analogical mind: perspectives from cognitive science. The MIT Press/Bradford Book, Cambridge

35. Holyoak KJ, Thagard R (1995) Mental leaps: analogy in creative thought. MIT Press, Cambridge

36. Howard TJ, Dekoninck EA, Culley SJ (2010) The use of creative stimuli at early stages of industrial product innovation. Res Eng Des 21:263–274. doi:10.1007/s00163-010-0091-4

37. Jansson DG, Smith SM (1991) Design fixation. Des Stud 12:3–11. doi:10.1016/0142-694X(91)90003-F

38. Jin Y, Li W, Lu SCY (2005) A hierarchical co-evolutionary approach to conceptual design. CIRP Ann Manuf Tech 54:155–158. doi: 10.1016/S0007-8506(07)60072-9

39. Jones JC (1984) A method of systematic design. In: Cross N (ed) Developments in design methodology. Wiley, New York

40. Kan JWT, Gero JS (2009) Using the FBS ontology to capture semantic design information in design protocol studies. In: McDonnell J, Lloyd P (eds) About: designing. Analysing design meetings. CRC Press, Leiden

41. Liikkanen LA, Perttula M (2009) Exploring problem decomposition in conceptual design among novice designers. Des Stud 30:38–59. doi:10.1016/j.destud.2008.07.003

42. Linsey JS, Wood KL, Markman AB (2008) Increasing innovation: presentation and evaluation of the Word tree design-by-analogy method. Proceedings of the 2008 ASME design theory and methodology conference. New York, 3–6 August

43. Liu YC, Chakrabarti A, Bligh T (2003) Towards an 'ideal' approach for concept generation. Des Stud 24:341–355. doi:10.1016/S0142-694X(03)00003-6

44. Luckman J (1984) An approach to the management of design. In: Cross N (ed) Developments in design methodology. Wiley, New York

45. March L (1984) The logic of design. In: N Cross (ed) Developments in design methodology. Wiley, New York

46. Nakakoji K, Yamamoto Y (2001) What does the representation talk-back to you? Knowl Base Sys 14:449–453. doi:10.1016/S0950-7051(01)00139-3

47. Osborn AF (1957) Applied imagination: principles and procedures of creative problem solving. Charles Scribner's Sons, New York

48. Oxman R (2002) The thinking eye: visual re-cognition in design emergence. Des Stud 23:135–164. doi:10.1016/S0142-694X(01)00026-6

49. Pahl G, Beitz W (1995) Engineering design: systematic approach. Springer, Berlin

50. Parka H, Sona JS, Leeb KH (2008) Design evaluation of digital consumer products using virtual reality-based functional behaviour simulation. J Eng Des 19:359–375. doi:10.1080/09544820701474129

51. Perttula M, Sipilä P (2007) The idea exposure paradigm in design idea generation. J Eng Des 18:93–102. doi:10.1080/09544820600679679

52. Pugh S (1991) Total design: integrated methods for successful product engineering. Addison-Wesley, Reading

53. Rahimian FP, Ibrahim R (2011) Impacts of VR 3D sketching on novice designers' spatial cognition in collaborative conceptual architectural design. Des Stud 32:255–291. doi:10.1016/j.destud.2010.10.003

54. Runco MA, Pritzker SR (1999) Encyclopedia of Creativity. Academic Press, San Diego

55. Sarkar P, Chakrabarti A (2011) Assessing design creativity. Des Stud 32:348–383. doi:10.1016/j.destud.2011.01.002

56. Sarkar S, Dong A, Gero JS (2010) Learning symbolic formulations in design: syntax, semantics, knowledge reification. AI EDAM 24:63–85. doi:10.1017/S0890060409990175

57. Schön DA (1983) The reflective practitioner: how professionals think in action. Temple Smith, London

58. Schön DA (1987) Educating the reflective practitioner: toward a new design for teaching and learning in the professions. Jossey-Bass, San Francisco

59. Shah JJ, Kulkarni SV, Vargas-Hernandez N (2000) Evaluation of idea generation methods for conceptual design: effectiveness metrics and design of experiments. J Mech Des 22:377–384. doi:10.1115/1.1315592

60. Shah JJ, Vargas-Hernandez N, Smith SM (2003) Metrics for measuring ideation effectiveness. Des Stud 24:111–134. doi:10.1016/S0142-694X(02)00034-0

61. Shai O, Reich Y, Hatchuel A, Subrahmanian E (2009) Creativity theories and scientific discovery: a study of C-K theory and infused design. In: Proceedings of the 17th international conference on engineering design. Stanford, CA, 24–27 August

62. Simon HA (1973) The structure of ill-structured problems. Artif Intell 4:181–200. doi:10.1016/0004-3702(73)90011-8

63. Sternberg RJ, Lubart T (1999) The concept of creativity: prospects and paradigms. In: Sternberg RJ (ed) Handbook of creativity. Cambridge University Press, Cambridge

64. Suh NP (1990) The principles of design. Oxford University Press, New York

65. Tovey M (1989) Drawing and CAD in industrial design. Des Stud 10:24–39. doi:10.1016/0142-694X(89)90022-7

66. Ulrich KT, Eppinger SD (2008) Product design and development, 4th edn. McGraw-Hill Higher Education, Boston

67. van der Lught R (2002) Functions of sketching in design idea generation meetings. In: Proceedings of the 4th conference on creativity & cognition. Loughborough, UK, 13–16 October, pp. 72–79. doi: 10.1145/581710.581723

68. Vargas-Hernandez N, Shar JJ, Smith SM (2010) Understanding design ideation mechanisms through multilevel aligned empirical studies. Des Stud 31:382–410. doi:10.1016/j.destud.2010.04.001

69. Vidal R, Mulet E, Gómez-Senent E (2004) Effectiveness of the means of expression in creative problem-solving in design groups. J Eng Des 15:285–298. doi:10.1080/09544820410001697587
70. Visser W (1996) Two functions of analogical reasoning in design: a cognitive-psychology approach. Des Stud 17:417–434. doi:10.1016/S0142-694X(96)00020-8
71. von Hippel E (1978) Successful industrial products from customer ideas. J Mark 42:39–49
72. Weisberg RW (1993) Creativity: beyond the myth of genius. WH Freeman and Co, New York
73. Yang MC (2010) Observations on concept generation and sketching in engineering design. Res Eng Des 20:1–11. doi:10.1007/s00163-008-0055-0

Chapter 3
Design Competence for Design Creativity

Abstract In this chapter, we discuss *design competence* which not only underlies the *concept generation* deep in the mind but also drives it. *Design competence* is defined as that which contains the basis and ability of the *concept generation* for *design creativity*. *Design competence* is categorized into three types: competence to inspire the motivation from inside a thought space, competence to abstract the concepts, and competence to control the *back-and-forth issue*. The former two competences are related to space issues—that is, the first one is a horizontal issue and the second, a vertical issue—whereas the latter competence is related to the issue of time.

3.1 Design Competence

In the previous chapter, *design creativity* was defined as the degree to which an *ideal* is conceptualized, and *concept generation* was defined as the process of *composing* a desirable concept towards the *future*, by focusing on the *inner sense–driven phase*. Following these definitions, ***design competence*** is defined as that which involves the 'basis' and 'ability' of the *concept generation* for *design creativity*.

3.2 Types of Design Competence

Design competence can be categorized into the following three types:

(1) Competence to inspire the motivation from inside a thought space,
(2) Competence to abstract the concepts, and
(3) Competence to control the *back-and-forth issue*.

(1) and (2) are related to space issues—that is, (1) is a horizontal issue and (2) is a vertical issue—whereas (3) is related to the issue of time.

T. Taura and Y. Nagai, *Concept Generation for Design Creativity*,
DOI: 10.1007/978-1-4471-4081-8_3, © Springer-Verlag London 2013

3.2.1 Competence to Inspire the Motivation from Inside a Thought Space

This competence determines and forms the boundary between the inside and outside of a thought space, as follows.

First, we should discuss the subissue regarding the direction from which the boundary is determined, that is, whether it is from the inside or outside. 'Autopoiesis', as applied to an organization, explains that boundaries will be determined from the inside [5]. On the basis of autopoiesis, Winograd and Flores [15] introduce the framework of a network system which is formed in a topological manner (i.e. autonomy). Winograd [14] asserts the importance of software engineering in the planning of an interactive system as a formed network system. On the other hand, the process of creating art can be viewed as a self-referential process or a self-recognition process, because during the creative process, it is impossible to separate the artist from the created work [4]. These are thought-provoking ideas that arise from the inside, and we suppose that the boundary of the thought space of design can be determined from the inside [11].

Next, we discuss a subissue regarding the motivation of *concept generation*. In order to capture the very essence of the *design competence*, it is necessary to focus on the motivation to drive the *concept generation*. Motivation has been considered by psychologists to be an important factor for creativity. It has been reported that highly creative work is produced by those who have strong 'intrinsic motivation' to engage in an activity [1, 2]. Whether the motivation is intrinsic or extrinsic is a topic of discussion. An extrinsic motivation is a stimulus from the outside (i.e. from an external source, e.g. a reward) which leads to humans channelling all their activities towards a particular goal. An intrinsic motivation is the inner motivation (i.e. from an internal source) to do an activity for the inherent satisfaction of the activity itself [9]; it is responsible for human (personal) behaviour and spans from the bionic level, for example, 'hunger', to a higher cognitive level, such as an artist's 'flow' (a state of concentration or complete absorption in an activity) [3]. The intrinsic motivation is thought to play an important role in the *design competence*.

3.2.2 Competence to Abstract the Concepts

The generic meaning of 'abstract' is to extract particular properties from the concept. Generating a new concept by combining multiple concepts is a sophisticated activity [8, 13] (this point will be discussed in more detail in Chap. 4). For example, if we knew the two concepts of "snow" and "car", we could derive new concepts from them such as "white steam locomotive", "yellow steam locomotive", "white post box", and "yellow post box". This process of *concept generation* comprises the following three steps: obtaining more general concepts, such as "white coloured objects" and "moving objects"; combining them, such as "moving object with a white colour", "moving object whose colour is not white", "non-moving object with a white colour", and "non-moving object whose colour is not white"; and generating a new concept by specifying the combined general concepts. By performing these steps, we could then generate new concepts. In this book, such a general concept, obtained by extracting particular properties or features (function, attribute, etc.) of objects, is defined as an ***abstract concept***, on applying General Design Theory (GDT) [16]. For example, "white" is a description of an *abstract concept* regarding the attribute of 'snow'. On the other hand, ***entity concept*** is defined as a concept regarding an object itself. For example, "snow" is a description of an *entity concept*. In this book, the *entity concept* or *abstract concept* of aaa is described as "aaa".

On the other hand, there is another meaning of 'abstract'. This is the meaning used in art; for example, in the term 'abstract painting'. In this usage, abstract paintings are drawn neither from the attributes of objects nor from the simpler representation of the object [6]. An abstract painting usually represents a motif which involves intuitive imagination or an associated image to motivate the artist. Such paintings are perhaps conceived in the mind of the artist. Historically, Impressionism is the base of this abstraction. In paintings after Impressionism, two ways of abstraction have been addressed. One is to systematically form a picture on the basis of transformation by using certain rules. This was based on the decomposing method of Impressionism to represent a real view, especially in colours. For this method, the artists normally decided some rules of the transformation in advance, like the grammar of the paintings. Mondrian is a famous artist and a pioneer of abstract painting in this way. First, he represented pictures by reducing the real sights. He represented a tree without including details in the different processes of the line art. Afterwards, he favoured building the classification rules of the shapes and colours oriented towards the basic construction of nature. For example, he divided the sight into three colours (blue, red, and yellow), and two directions (bars and poles) in black. Not only geometrical pictures but also organic pictures, like 'art informel', are developed from representative paintings because both initiated abstraction from the sight of the real world.

The other method of abstract painting is to represent the meta-reality (psychological reality). Mondrian's later works, representing his dream of life, were much closer to this method. Cézanne is another pioneer of abstract painting. He is also famous for abstract paintings which simplify basic forms (cubes and spheres); however, he chased the same motif of his artistic theme. He paid attention to the abstracted theme rather than the abstracting process. Thus, his picture is expected to be richer than the real. Picasso, who developed Cubism [7], identified the role of abstraction of painting from Cézanne's chased theme as one expressing the reality of a motif itself. For example, he painted an abstract image of a guitar as still life, and this image expressed a musical atmosphere rather than the shape of the guitar. We can say his pictures were derived from the meta-reality of his mind, which can be his psychological real world. Kandinsky also represented pictures using this method. His painting lyrically expressed illusions. There were no sights similar to

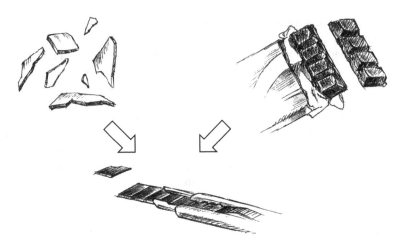

Fig. 3.1 Art knife designed by combining two concepts: broken glass and chocolate segments

meta-reality in the real world, but it exists in his mind. To date, it is far more difficult to create pictures like those of Picasso or Kandinsky using a computer.

The competence to abstract the concepts in both the above ways is thought to play an important role in *design competence*.

3.2.3 Competence to Control the Back-and-Forth Issue

During the process of *concept generation*, we often make decisions that can be evaluated only after the process has progressed for a while. We call this issue a *back-and-forth issue in Design* [10] and believe that controlling this issue composes an element of the *design competence*.

For example, let us consider the process of combining two concepts. Although this process is regarded as the most essential process in generating a new concept from the existing ones as mentioned above, it is extremely difficult to select the appropriate concepts to be combined before generating because the appropriateness of these concepts can only be evaluated after they have been combined and the generated outcome has been evaluated. From an empirical viewpoint, the invention of the art knife—the first snap-off blade cutter—is a good example (Fig. 3.1). It is said that the inspiration for this incredible idea came from the combination of two concepts: "chocolate segments which can be broken off" and "sharp edges of broken glass" (OLFA[1]). Although this invention is rather attractive, the issue of

---

[1] OLFA Corporation. World's first snap-off blade cutter. http://www.olfa.co.jp/en/contents/ cutter/birth.html. Accessed 17 October 2011.

Fig. 3.2 Path of the beam through a reflection on a mirror

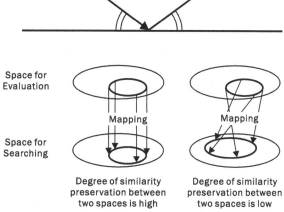

Fig. 3.3 Preservation of the similarity between two spaces (evaluation space and searching space)

Space for Evaluation

Mapping   Mapping

Space for Searching

Degree of similarity preservation between two spaces is high

Degree of similarity preservation between two spaces is low

focusing on the chocolate remains unanswered. In other words, why is the chocolate focused on? Generally, the chocolate is unconcerned with the utility knife.

In certain cases, the *back-and-forth issue* can take the form of a spatial issue. For example, consider the situation where we attempt to identify a beam of light that passes through a reflection on a mirror (Fig. 3.2). If we predict the beam's path on the basis of the knowledge that 'a beam of light travels along the path that takes the shortest time', we would be unable to evaluate whether or not a path takes the shortest time before the beam has virtually travelled.

However, if we apply the knowledge that 'the angle of incidence is equal to the angle of reflection', then it becomes possible to calculate the path of the light beam before we actually observe the travelling beam. In this case, the *back-and-forth issue* from the viewpoint of time is converted into a spatial issue.

General Design Theory provides a rigorous method in this area. In GDT, the design process is defined as a mapping from the function space—where the specification is described and searched candidates of solutions are evaluated—to the attribute space—where the shape or structure of a solution is searched. In this case, to effectively search for a solution (*entity concept*), it is necessary to determine an appropriate searching space; in particular, to determine the properties (attributes) which are used to search for the solution. This determination is a *back-and-forth issue*, since the appropriateness of the searching space can only be evaluated after searched candidates of solutions are evaluated. For this determination, it is expected that the introduction of a metric into the design spaces (evaluation space and searching space) and the preservation of the similarity between these two spaces, make it possible to effectively search for a solution [12]. In other words, if two *entity concepts* are close to each other in the searching space, under the condition that the same *entity concepts* are close to each other in the evaluation space, then the search for a solution may be effective (Fig. 3.3).

This rule is valid only when the solution is searched for using a neighbourhood search method.

Taura confirms the above described method of converting the *back-and-forth issue* into a spatial issue by applying it to the function decomposition process in design [10].

Although we have yet to clarify the mechanism to control the *back-and-forth issue*, potential designers are thought to have the competence to control it in an implicit manner.

References

1. Amabile TA (1988) A model of creativity and innovation in organizations. In: Staw BM, Cummings LL (eds) Research in organizational behavior, vol 10. JAI Press, Greenwich, pp 123–167
2. Amabile TA (1996) Creativity in context: update to the social psychology of creativity. Westview Press, New York
3. Csikszentmihalyi M (1988) Motivation and creativity, toward a synthesis of structural and energistic approaches to cognition. New Ideas Psychol 6:159–176. doi:10.1016/0732-118X(88)90001-3
4. Hass L (2008) Merleau-Ponty's philosophy. Indiana University Press, Bloomington
5. Maturana HR, Varela FJ (1980) Autopoiesis and cognition: the realization of the living. D Reidel, Dordrecht
6. Nagai Y, Taura T (2009) Design motifs: abstraction driven creativity—a paradigm for an ideal design. Special Issue Jpn Soc Sci Des 16–2:13–20
7. Read H (1974) A concise history of modern painting. Thames and Hudson, London
8. Rothenberg A (1979) The emerging goddess: the creative process in art, science, and other fields. University of Chicago Press, Chicago
9. Ryan R, Deci E (2000) Self-determination theory and the facilitation of intrinsic motivation, social development, and well-being. American Psychologist 55:68–78
10. Taura T (2008) A solution to the back and forth problem in the design space forming process: a method to convert time issue to space issue. Artifact 2:27–35
11. Taura T, Nagai Y (2009) Design creativity: integration of design insight and design outsight. Special Issue Jpn Soc Sci Des 16–2:55–60
12. Taura T, Yoshikawa H (1992) A metric space for intelligent CAD. In: Brown DC, Waldron MB, Yoshikawa H (eds) Intelligent computer aided design. North-Holland, Amsterdam
13. Ward TB, Smith SM, Vaid J (1997) Creative thought: an investigation of conceptual structures and processes. APA Book, Washington
14. Winograd T (1996) Bringing design to software. ACM Press, New York
15. Winograd T, Flores F (1986) Understanding computers and cognition: a new foundation for design. Ablex Publishing, Norwood
16. Yoshikawa H (1981) General design theory and a CAD System. In: Sata T, Warman EA (eds) Man-machine communication in CAD/CAM. North-Holland, Amsterdam

Chapter 4
Theory of Concept Generation

Abstract In this chapter, a systematized theory of *concept generation* is developed. The theory classifies the *concept generation* into the following two types: *first-order concept generation*, which is based on the similarity-recognition process, and *high-order concept generation*, which is based on the dissimilarity-recognition process. The close investigation suggests that *first-order concept generation* is related to the *problem–driven phase* and *high-order concept generation* is related to the *inner sense–driven phase*. Furthermore, the methods to select the appropriate *base concepts* are discussed as follows. In *first-order concept generation*, a cue is thought to be obtained in the process of analysing the existing products, while in *high-order concept generation*, a cue can be captured in the framework of the *back-and-forth issue*.

4.1 Abstract Concept Versus Entity Concept

Abstract concepts and *entity concepts* have been defined in Chap. 3. An *abstract concept* is a concept which is obtained by extracting particular properties or features (function, attribute, etc.) of objects, whereas an *entity concept* is a concept regarding an object itself. Accordingly, the process of obtaining the *abstract concept* is considered the process of extracting a number of common properties or features (attributes or function) from a number of *entity concepts*. Using set theory, the *entity concept* is modelled as an element, and the *abstract concept* is modelled as a class (subset of elements), since the *entity concept* which has the same attribute or function (*abstract concept*) can be gathered in the subset of the *abstract concept* (Fig. 4.1).

Here, we would like to affirm that discussion on the *concept generation* should deal with concepts which do not exist yet in the real world. That is, we will discuss the theory for the generation process of concepts by identifying the difference

Fig. 4.1 Entity concept and
abstract concept

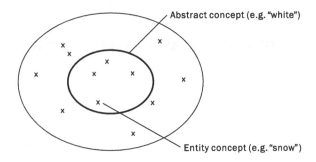

Abstract concept (e.g. "white")

Entity concept (e.g. "snow")

between 'the concepts which do not exist yet' and 'the existing or existed concepts' in the real world. However, as mentioned in Chap. 1, we assume the situation where a new concept is generated with a certain basis and by referring to the existing concepts (these concepts will be discussed as *base concepts* in the following sections).

The question to now be addressed is which type of concept—*abstract* or *entity*—should be discussed. When we identify the purpose of *concept generation* to be the generation of a new *entity concept* (product), we need to discuss the *entity concept*; however, when we identify the purpose of *concept generation* to be the generation of a new attribute or function, we must discuss the *abstract concept*. In this book, we primarily discuss the *entity concept* because the object of design is considered an entity. However, at the same time, we take the *abstract concept* into account since the *abstract concept* plays the important role of generating the *entity concept*; in addition, the *abstract concept* itself can be an object for *concept generation*. In this book, what is referred to as 'concept' is the *entity concept*.

4.2 Classification of Concept Generation

We attempt to develop a systematized theory for *concept generation* by referring to the models of metaphor, abduction, and the operation of multiple *abstract concepts* by referring to General Design Theory (GDT) [27], especially from the viewpoint of recognition of the similarity and dissimilarity between the concepts.

4.2.1 First-Order Concept Generation Based on the Similarity-Recognition Process

A metaphor is a rhetorical figure representing one concept by way of another concept, which is considered similar to the concept to be represented. The roles of metaphors have also been investigated in design [4, 7, 10, 11, 26]. If A is a concept to be represented and B is a concept to be assimilated to, the relationship between A and B

Fig. 4.2 Structure of a metaphor

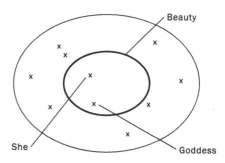

Fig. 4.3 Swan chair (sketched by the author)

is expressed as 'A is B'. For example, let us consider the expression 'she is a goddess'. In this case, the common features between 'she' and 'goddess' are implicitly represented (Fig. 4.2). Both 'she' and 'goddess' belong to the same *abstract concept* which was formed from the *commonality* between 'she' and 'goddess', for example, "beauty".

Metaphors are often used in design. For example, Fig. 4.3 is an illustration of a famous chair called the Swan Chair. This chair was created by the Danish designer Arne Jacobsen for a hotel, and it remains very popular. As its name conveys, the Swan Chair is a chair which approximates the shape of a swan in order to convey the quality of a swan (i.e. elegance) and can be represented by the framework for a metaphor, such as 'the chair is a swan'. Although it sounds similar to the previous sentence ('she is a goddess'), the roles present in the metaphors are different. When a person says 'she is a goddess', she already exists; in contrast, the sentence 'the chair is a swan' presents a more complicated situation. If a person says this when he or she looks at the chair (Fig. 4.4), it refers to an existing chair in the same vein as 'she is a goddess'. On the other hand, if a person says this while he or she is designing the chair, then the chair is a concept which does not yet exist. In this situation, 'the new chair which is to be designed is a swan' is a more

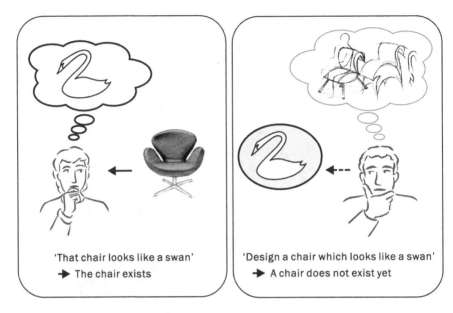

'That chair looks like a swan'
➤ The chair exists

'Design a chair which looks like a swan'
➤ A chair does not exist yet

Fig. 4.4 The two types of metaphors in design

suitable representation. In such a case, the similarity (implicit common features; e.g. elegant shape) between the chair (which does not yet exist but will be designed) and a swan play an important role. On the basis of the above considerations, we can say that, in the design process using metaphors, a new concept will be generated which will be similar to the concept to be assimilated to. A new concept of a chair is generated (referring to a swan or imitating a swan) on the basis of the similarity between the chair to be designed and a swan.

In our previous study, the process of generating a new concept was investigated using the example of a 'snow tomato' [16]. In this chapter, we also discuss the nature of *concept generation* using this example. From the concepts of 'snow tomato', it is easy to generate the new concept of "white tomato", wherein the property of colour is the focus. The generation process of this new concept can be explained using the metaphor framework, such as 'the tomato to be designed is snow'. This representation involves the notion that "tomato" is related to "snow"; both concepts are not closely related to each other in the usual way.

On the other hand, *concept generation* has been discussed within the framework of reasoning. The findings of previous studies indicate that abduction is the most adequate representation of the characteristics of design among the three reasoning types: deduction, induction, and abduction [6]. This is because design is understood to be the process of developing a hypothesis. In this chapter, we show that the two representations (metaphor and abduction) result in the same meaning.

The process of *concept generation* for a Swan Chair can also be formulated as a process of abduction (Fig. 4.5). This formulation can be interpreted as follows.

Fig. 4.5 Concept generation
by abduction

$$\forall x(swan(x) \rightarrow elegant\_shape(x)) \quad (1)$$

$$elegant\_shape(chair\_to\_be\_designed) \quad (2)$$

$$swan(chair\_to\_be\_designed) \quad (3)$$

Formula (1) shows that the *abstract concept* (subset) of "swan" is included in the *abstract concept* (subset) of "elegant shape". Formula (2) shows that "chair to be designed" (element) belongs to the *abstract concept* (subset) of "elegant shape". Consequently, "chair to be designed" and the *abstract concept* of "swan" are connected by formula (3), whereas "chair to be designed" does not belong to "swan". This formulation is explained as the process in which "chair to be designed" is related to "swan" because of the common property of "elegant shape". The two concepts are connected on the basis of a similar property; in other words, a novel concept is generated such that a similar property is preserved. This explanation involves nearly the same structure as that of a metaphor (the slight difference is owing to the type of concept: an *entity concept* in a metaphor and an *abstract concept* in abduction. Actually, the description of an object can be an *abstract concept* as well as an *entity concept*. For example, "car" can be an *abstract concept* which includes vehicles which have the properties of 'car', as well as an *entity concept*).

From the above considerations, one finds that, in the very early stage of the design process which is represented as abduction or metaphor, either a new concept is related to an existing concept (although these are not related to each other in the usual way) owing to some similarity between them, or a new concept is generated such that the similarity is preserved. This relation can be called 'grouping'. "Swan Chair" (a new concept) and "swan" are in the same group. "White tomato" (a new concept) and "snow" are in the same group. On the basis of the grouping process, a new chair or a new tomato is generated.

The above consideration is consistent with Ito's discourse [14]. He states that the features of tacit knowledge proposed by Polanyi [18] and abduction are based on a process of synecdoche, which is the basic operation of metaphor and grouping. In addition, the KJ Method and other models of *concept generation* are types of groupings based on the similarity-recognition process [17, 21].

Although metaphor and abduction seem to be apparently different processes, we explained above that both processes have the same nature; that is, they are grouping processes based on the similarity-recognition process. On the basis of the considerations discussed in this section, we define the *abstract concept* generated by focusing on a similar property or feature (function or attribute) as a ***first-order abstract concept***, and the ***first-order concept generation*** as the process of generating a new concept on the basis of the *first-order abstract concept* (Fig. 4.6).

In a *first-order concept generation*, a new concept is generated by mapping a common property or feature (*first-order abstract concept*) to an existing concept. For example, the new concept of "Swan Chair" by Jacobsen was generated by transferring the shape of "swan" to the shape of an existing chair. The new

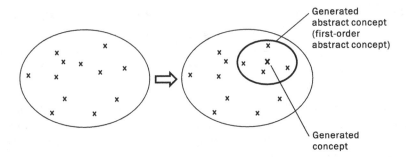

Fig. 4.6 First-order concept generation

concept of "white tomato" is generated by transferring the colour of "snow" (white) to the colour of "tomato". This transferring process is defined as ***property mapping***. Hereafter, the existing concept and the concept whose property is mapped to the existing concept are called ***base concepts*** in order to ensure consistency with the other processes in this book, although the former is conventionally called the 'target domain', and the latter, the 'source domain'. In *property mapping*, various properties or features (similarities) may be transferred from the *base concept*, in addition to the salient property or feature which is first focused on, which are addressed in the previous discourses on metaphor [2, 12, 13, 15, 25].

As a conclusion to this subsection, generating a new concept using metaphors or abduction is understood to be *property mapping* based on the similarity-recognition process.

However, a *first-order concept generation* is useful only to create a subspecies of an existing object, since it cannot extend beyond the category of the existing product. For example, in the case of "Swan Chair", a new concept belonging to the category beyond the chair cannot be generated while a creative chair is expected to be generated.

Another important issue to be identified is that the *property mapping* is effective for only explaining *the concept generation* in an 'ex post facto' manner. For example, in the case of "Swan Chair", the issue of how or why a swan was chosen remains unanswered. Generally, swans have nothing to do with chairs. We describe this dilemma in a later section of this chapter.

4.2.2 High-Order Concept Generation Based on the Dissimilarity-Recognition Process

We would like to discuss another method of generating a new concept. Let us consider the example of a 'snow tomato' again. By using the *property mapping* of a *first-order concept generation*, we can obtain the new concept of "white tomato". However, the generated concept is a kind of 'tomato' and is not a concept

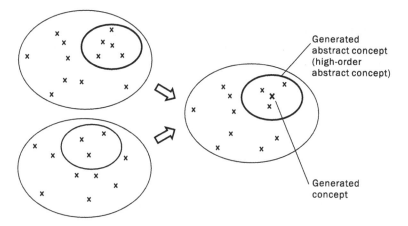

Fig. 4.7 High-order concept generation

beyond the category of 'tomato'. On the other hand, if we perform the following steps of operating multiple *abstract concepts,* a new concept of "powdered ketchup", which is used like powdered cheese and thought to extend beyond the categories of 'tomato' and 'snow', can be generated: we obtain the *abstract concepts* of "flavouring" and "snowflake" from "tomato" and "snow", respectively, and combine them to acquire "flavouring with snowflake", thus generating the new concept of "powdered ketchup" by specifying the *abstract concept* of "flavouring with snowflake". By performing these steps, we could then generate another class of new concepts rather than *first-order concept generation.* On the basis of the above considerations, we define the *abstract concept* generated by combining multiple abstract concepts (hereafter, the *entity concepts* from which these *abstract concepts* are generated are also referred to as **base concepts**) as a **high-order abstract concept**, and the **high-order concept generation** as the process of generating a new concept on the basis of the *high-order abstract concept* (Fig. 4.7).

Here, the relationship between the properties or features is crucial in determining whether the generated concept may extend beyond the existing categories or not. With regard to the example of 'snow tomato', when the property of "red" of "tomato" is replaced with the property of "white" of "snow", then the concept of "white tomato" is obtained. In this case, the relation between "white" and "red" is called an *alignable difference* and pointed out to be related to 'similarity' in the field of cognitive science. In the *property mapping*, these alignable properties are replaced. On the other hand, in the case of "powdered ketchup", the relationship between the extracted properties of "flavouring" and "snowflake" is called a *nonalignable difference* and pointed out to be a kind of 'dissimilarity'. In this case, "flavouring" and "snowflake" cannot be replaced, since they do not align in the same dimension (these points will be further discussed in Chap. 5.).

The above considerations suggest that a *high-order concept generation* is understood to have its basis on the dissimilarity-recognition process. In other words, the dissimilarity plays an important role in generating an *innovative* concept which extends beyond the existing categories. In this book, hereafter, the term *innovative* is used to denote the notion of 'extending beyond the existing categories'.

High-order abstract concepts are classified into two types. Generally, in the relation between concepts, it has been revealed that there are two types of relations—taxonomical and thematic [22]. The former is a relation that represents a physical resemblance between the concepts, and the latter represents the relation between the concepts through a thematic scene. With respect to the example of an 'apple' and an 'orange', the relation which focuses on the shape (round) is a taxonomical relation. On the other hand, with respect to the example of an 'apple' and a 'knife', the relation which focuses on the scene in which 'an apple is cut by a knife' is a *thematic relation*.

According to the former relation, *high-order abstract concepts* can be interpreted as *abstract concepts* involving an *innovative* concept which inherits partial properties from both the two *base concepts* but is different from the two *base concepts*. We define the *concept generation* based on this process as **concept blending**, echoing Fauconnier's discourse. In the cognitive linguistic study, Fauconnier [8] proposes 'the blended' space on the basis of his mental space theory. In conceptual blending, two input concepts yield a third concept which inherits partial features from the input concepts and has emergent features of its own [9]. The concept of "powdered ketchup" is an example of *concept blending*. This new concept is understood to be generated by blending the different properties of snow (*abstract concept*: snowflake) and tomato (*abstract concept*: flavouring).

On the other hand, according to the latter relation between the two *base concepts*, *high-order abstract concepts* can be interpreted as *abstract concepts* involving an *innovative* concept which is generated from the thematic scenes (situations, roles, etc.) of the *base concepts*. We define the *concept generation* based on the *thematic relation* as **concept integration in thematic relation**. In design, the outcomes (the designed products) must be meaningful to people. Therefore, the designer must carefully consider not only the attributes of the designed product (shape, material, etc.) but also its function and interface; in other words, careful consideration of the human aspect is important. Accordingly, integration of the *base concepts* in the *thematic relation* is expected to play an important role in *concept generation* (this point will be discussed in more detail in Chap. 6). With respect to the example of a 'snow tomato', the new concept of "refrigerator which can humidify the food in it" is generated from the scene where "tomato" is stored in "snow".

The notion of combining multiple *abstract concepts* on the basis of the 'dissimilarity-recognition process' of *high-order concept generation* mentioned above is shared with GDT. In GDT, AXIOM 3 is strongly related to the manipulation of multiple *abstract concepts*.

(AXIOM 3) (Axiom of operation) The set of *abstract concepts* is a topology of the set of the *entity concept*.

Here, 'topology' refers to mathematical topology and has the following properties.

1. The entire *entity concept*, that is, the universal set of the *entity concept*, is an *abstract concept*. Inversely, the null set of the *entity concept* is an *abstract concept*.
2. By definition, an *abstract concept* is a subset of the *entity concept* set. The intersection of an *abstract concept* and another *abstract concept* is also an *abstract concept*.
3. A union of *abstract concepts* is also an *abstract concept*.

AXIOM 3 gives a high-operability to the *abstract concepts*. In particular, the argument that a new *abstract concept* is obtained as the intersection of two *abstract concepts* is similar to the process in the *high-order concept generation*. Regarding the role of topology in design, further discussion has been conducted [3, 19, 24]. In GDT, some theorems have been deduced. From among them, we introduce Theorem 1.

(Theorem 1) The ideal knowledge is the Hausdorff space.

This theorem is deduced from the fact that ***ideal knowledge*** is defined as a knowledge in which all the elements of the entity set are known and in which each element can be described by *abstract concepts* without ambiguity. A Hausdorff space is a separated space, which implies the notion of differentiation. This discussion suggests that GDT, which operates the concept on mathematical topology and defines the *ideal knowledge* as a separated space, is strongly related with the notion of *high-order concept generation*.

As mentioned above, *high-order concept generation* is expected to generate an *innovative* concept which extends beyond the existing categories. However, we must note that the type of *concept generation* is not absolutely determined owing to the newly generated concept itself; instead, it is relatively dependent on the *base concepts* from which the new concept has stemmed. For example, in the case of 'snow tomato', the new concept of "powdered ketchup" can also be obtained by the *property mapping*, where the property of "powdered" is transferred to "ketchup" when "powdered cheese" and "ketchup" are recognized as the *base concepts*. Moreover, the new concept of "refrigerator which can humidify the food in it" can also be obtained by *property mapping*, where the property of "humidifying" is transferred to "refrigerator" when "humidifier" and "refrigerator" are recognized as the *base concepts*. Accordingly, perhaps one can consider *concept blending* or *concept integration in thematic relation* as an extension of the *property mapping*.

Indeed, *property mapping*, *concept blending*, and *concept integration in thematic relation* can be identified as the same process to find the intersection of *abstract concepts* in set theory, while the difference among them results in the interpretation of the intersection. However, the meanings of the two processes (*first-order concept generation* and *high-order concept generation*) are quite

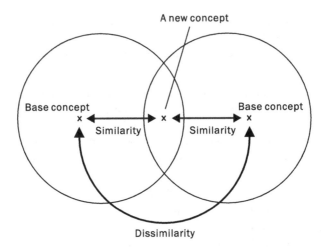

Fig. 4.8 Similarity and dissimilarity in concept generation

different. The generated concepts differ as to whether they should be a subspecies of the existing product or may extend beyond the existing category, and as to whether they are based on the 'similarity-recognition process' (which denotes that concepts belong to the same *abstract concept*) or 'dissimilarity-recognition process' (which denotes that concepts belong to different *abstract concepts*). The relation between these processes is illustrated in Fig. 4.8. The difference in the recognition process between *first-order concept generation* and *high-order concept generation* is discussed again by referring to the linguistic interpretation process in Chap. 5.

On the other hand, in the same way as *property mapping*, the model of manipulating multiple *abstract concepts* is also effective for explaining *concept generation* through only in an 'ex post facto' manner. We can confirm the feasibility of manipulating the *abstract concepts* only from the prearranged *base concepts*, and describe *concept generation* as a *high-order concept generation* in an 'ex post facto' manner; further, it is impossible to answer the following question: Why were those specific *base concepts* focused on from the beginning? We describe this issue in the following section.

4.3 How to Select the Appropriate Base Concepts for Concept Generation

In the previous sections, questions on how we select the appropriate concept to be related before the *property mapping* (*first-order concept generation*) and how we select the appropriate *base concepts* before the process of manipulating multiple *abstract concepts (high-order concept generation)* remain unanswered. In this section, we discuss these issues.

First, we discuss the *property mapping (first-order concept generation)*. In the *property mapping*, the concept to be related is assumed to be identified by analysing the properties or features of an existing concept. For example, it is imagined that the idea of a Swan Chair might have emerged during the designer's experience of looking at a 'swan'—that is, a designer who had seriously thought about the *problems* of existing chairs (elegant shape) back then. The designer is assumed to have noticed the shape of the 'swan' while he was searching for an elegant shape. Accordingly, in the *property mapping (first-order concept generation)*, a cue for identifying the concept which is to be related is thought to be obtained in the process of analysing the existing *problem*. In other words, the *base concepts* should be selected such that the *problem* can be solved. We have implemented this idea into our functional reasoning method [20].

Next, we discuss *concept blending* and *concept integration in thematic relation (high-order concept generation)*. Taura [23] discusses the *back-and-forth issue*, as described in Chap. 3. His findings can be applied to the issue of how to select the *base concepts*. The method is to form a space to select the *base concepts*, so that the *base concepts* close to each other, in the space where the *base concepts* are selected (searched), are also close to each other in the space where the outcomes, generated from the same *base concepts,* are evaluated. Regarding the process where a new concept is generated from *base concepts*, the following findings were obtained (this point will also be discussed in Chap. 8).

1. In the case where the distance between the *base concepts* is high, the designed outcome with a high score for originality is obtained.
2. In the case where the *base concepts* are ones from which many other concepts can be associated, the designed outcome with a high score for originality is obtained.

By accumulating further findings, it is possible to select appropriate *base concepts*.

4.4 Role of High-Order Concept Generation

4.4.1 Relationships between the Phases of Concept Generation and the Types of Concept Generation

As described above, an important part of *property mapping (first-order concept generation)* is situated in the analysis of existing products. This consideration indicates that *property mapping* is closely tied to the *problem–driven phase*. Indeed, many cases of *property mapping* through problem analysis have been reported. For example, in the case of 'snow tomato', the new concept of "powdered ketchup" can

Table 4.1 Relationship between the phases of concept generation, types of concept generation, and the recognition process

Phases of concept generation	Type of concept generation	Recognition process
Problem–driven phase	Property mapping (first-order)	Similarity
Inner sense–driven phase	Concept blending and concept integration in thematic relation (high-order)	Dissimilarity

be obtained by *property mapping* when the "powdered cheese" and "ketchup" are the *base concepts*; this *property mapping* can be understood to be tied to the problem analysis: bottled ketchup is not easily handled. Moreover, the new concept of "refrigerator which can humidify the food in it" can be obtained by *property mapping* when the "humidifier" and "refrigerator" are the *base concepts*; this *property mapping* can be understood to be tied to the problem analysis: a refrigerator dries up the food inside it. Furthermore, an interesting case was reported wherein a subject in an experimental study analysed a *problem* during a design task through the use of mapping. The idea of a structure which involves opening and shutting was inspired by the rotation of a garage door and mapped into part of a stationery design [1, 5].

In *concept blending* and *concept integration in thematic relation* (*high-order concept generation*), their importance lies in the process of *composing* an assembly of elements (properties, etc.). This suggests that *high-order concept generation* involves the notion of *composing*. There may be an uncountable number of *composing* processes for selecting the appropriate *base concepts*, extracting the properties from the *base concept*, combining the *abstract concepts*, and elaborating on a more specific new concept from the obtained *abstract concept*. To perform these *composing* processes appropriately, the designer's *inner sense* seems to play an important role. If the new concept can be obtained only by analysing the current situation, it would be better if *first-order concept generation* was used. A motivation for employing *high-order concept generation* may be derived from the designer's wish to pursue a more *innovative* concept which is beyond the existing categories. These considerations indicate that *high-order concept generation* is closely related to the *inner sense–driven phase*, which involves the notion of *composing* with the *inner sense* for pursuing an *ideal*, since extending beyond the existing categories is assumed to share a common notion with pursuing an *ideal*.

From the above considerations, the relationships between the phases of *concept generation* and types of *concept generation* are summarized in Table 4.1. However, this relationship is not identified exclusively. For example, even in *first-order concept generation*, the *inner sense–driven phase* is included in it as its element since the *property mapping* cannot be completed by only analysing the situation. For instance, in order to find an appropriate *base concept*, a designer's *inner sense* is thought also to play an important role.

4.4.2 Meanings of High-Order Concept Generation

Although *high-order concept generation* has been considered essential to *concept generation*, it is difficult to detect it independently in conventional design activity. A reason for this is assumed to be that its strong point—its ability to generate a new concept which extends beyond the existing categories—is also its weak point in conventional design. Various ideas that extend beyond the existing categories can be obtained by *concept blending* or *concept integration in thematic relation*; however, this method alone is not applicable to the conventional design because we cannot predict the category of the ideas to be obtained in advance. When we design a chair in a conventional design, the result of the design must be a chair. If this is the case, how is *high-order concept generation* expected to contribute to real-world design? We would like to note the finding that the expansion of the thought space during the very early stage of design leads to a highly creative designed outcome (this point will be discussed in Chaps. 6 and 7). An important role of *high-order concept generation* is to expand the thought space on the basis of a designer's *inner sense*, at the very early stage of the design process. This role is expected to contribute to the highly *innovative* design which aims to create a highly *innovative* product extending beyond the existing category of products. This role is also expected to contribute to the conventional design in order to widen the designer's viewpoints so that the thought space is expanded at the very early stage of design even though the generated wide viewpoints are not directly related to the designed outcomes.

References

1. Ball LJ, Christensen BT (2009) Analogical reasoning and mental simulation in design: two strategies linked to uncertainty resolution. Des Stud 30:169–186. doi:10.1016/j.destud.2008.12.005
2. Black M (1979) More about metaphor. In: Ortony A (ed) Metaphor and thought. Cambridge University Press, New York
3. Braha D, Reich Y (2003) Topological structures for modeling engineering design processes. Res Eng Design 14:185–199. doi:10.1007/s00163-003-0035-3
4. Bucciarelli LL (2002) Between thought and object in engineering design. Des Stud 23:219–231. doi:10.1016/S0142-694X(01)00035-7
5. Christensen BT, Schunn CD (2007) The relationship of analogical distance to analogical function and preinventive structure: the case of engineering design. Mem Cognit 35:29–38
6. Coyne RD, Rosenman MA, Radford AD, Gero JS (1987) Innovation and creativity in knowledge-based CAD. In: Gero JS (ed) Expert systems in computer-aided design. North-Holland, Amsterdam

7. D'Souza NS (2010) The metaphor of an ensemble: design creativity as skill integration. In: Taura T, Nagai Y (eds) Design creativity 2010. Springer, London
8. Fauconnier G (1994) Mental spaces: aspects of meaning construction in natural language. Cambridge University Press, Cambridge, New York
9. Fauconnier G, Turner M (2002) The way we think: conceptual blending and the mind's hidden complexities. Basic Books, New York
10. Goldschmidt G, Litan Sever A (2011) Inspiring design ideas with texts. Des Stud 32:139–155. doi:10.1016/j.destud.2010.09.006
11. Hey J, Linsey J, Agogino AM, Wood KL (2008) Analogies and metaphors in creative design. Int J Eng 24:283–294
12. Indurkhya B (2006) Emergent representations, interaction theory and the cognitive force of metaphor. New Ideas Psychol 24:133–162. doi:10.1016/j.newideapsych.2006.07.004
13. Indurkhya B (2007) Rationality and reasoning with metaphors. New Ideas Psychol 25:16–36. doi:10.1016/j.newideapsych.2006.10.006
14. Ito M (1997) Tacit knowledge and knowledge emergence. In: Taura T, Koyama T, Ito M, Yoshikawa H (eds) The nature of technological knowledge. Tokyo University Press, Tokyo (in Japanese)
15. Lakoff J, Johnson M (1980) Metaphors we live by. University of Chicago Press, Chicago
16. Nagai Y, Taura T, Mukai F (2009) Concept blending and dissimilarity: factors for creative concept generation process. Des Stud 30:648–675. doi:10.1016/j.destud.2009.05.004
17. Oiwa H, Kawai K, Koyama M (1990) Idea processor and the KJ method. J Inform Process 13:44–48
18. Polanyi M (1966) The tacit dimension. Peter Smith, Gloucester
19. Reich Y (1995) A critical review of general design theory. Res Eng Des 7:1–18. doi:10.1007/BF01681909
20. Sakaguchi S, Tsumaya A, Yamamoto E, Taura T (2011) A method for selecting base functions for function blending in order to design functions. In: Proceedings of 18th International Conference on Engineering Design, Copenhagen, vol. 2, pp 73–86
21. Scupin R (1997) The KJ method: a technique for analyzing data derived from Japanese ethnology. Hum Organ 56:233–237
22. Shoben EJ, Gagne CL (1997) Thematic relation and the creation of combined concepts. In: Ward TB, Smith SM, Vaid J (eds) Creative thought: an investigation of conceptual structures and processes. APA Book, Washington
23. Taura T (2008) A solution to the back and forth problem in the design space forming process—a method to convert time issue to space issue. Artifact 2:27–35
24. Tomiyama T, Yoshikawa H (1985) Extended general design theory. In: Yoshikawa H, Warman EA (eds) Design theory for CAD. North-Holland, Amsterdam
25. Turbayne CM (1962) The myth of metaphor. Yale University Press, New Haven
26. Wang HH, Chan JH (2010) An approach to measuring metaphoricity of creative design. In: Taura T, Nagai Y (eds) Design creativity 2010. Springer, London
27. Yoshikawa H (1981) General design theory and a CAD System. In: Sata T, Warman EA (eds) Man-machine communication in CAD/CAM. North-Holland, Amsterdam

Chapter 5
Methods and Essence of Concept Generation

Abstract In this chapter, *concept generation* is systematized into more specific methods of *concept synthesis*: *property mapping*, *concept blending*, and *concept integration in thematic relation*. First, we show that these methods correspond to *property mapping*, *hybrid*, and *relation linking* in the field of linguistic studies, respectively. This correspondence not only makes it possible to compare *concept synthesis* with linguistic interpretation but also validates the classification of the three methods of *concept synthesis*. Next, we conduct an experiment to compare *concept synthesis* with the process of linguistic interpretation by focusing on the recognition types: *commonality*, *alignable difference*, and *nonalignable difference*. This experiment reveals that the main factors in the *concept synthesis* are *concept blending* and *nonalignable difference*. Moreover, we suggest that focusing on *nonalignable difference* is not a trait accumulated in subjects; rather, it occurs with regard to the design and interpretation processes.

5.1 Concept Synthesis

In Chap. 4, a systematized theory of *concept generation* was developed. In this chapter, more specific methods are systematized, by focusing on the process of combining the *base concepts* which are composed of two concepts. In this book, the process of combining the *base concepts* is termed **concept synthesis**. The advantage of this process is that it is the simplest and most essential process for generating a new concept from existing ones [5, 9].

Furthermore, this process is suitable because of the following two reasons.

The first reason is related to the empirical aspect. *Concept synthesis* is found in an actual field. The invention of the art knife—the first snap-off blade cutter—which was introduced in Chap. 3, is an appropriate example.

T. Taura and Y. Nagai, *Concept Generation for Design Creativity*,
DOI: 10.1007/978-1-4471-4081-8_5, © Springer-Verlag London 2013

	Property mapping	Concept blending	Concept integration in thematic relation
Example	"white tomato" 	"powdered ketchup" 	"humidifying refrigerator" 

Fig. 5.1 The three methods of concept synthesis ("tomato" and "snow")

The second reason is related to the methodological aspect. *Concept synthesis* can involve the three types of *concept generation* described in Chap. 4: *property mapping*, *concept blending*, and *concept integration in thematic relation*. *Property mapping* is a method of *first-order concept generation*, whereas *concept blending* and *concept integration in thematic relation* are methods of *high-order concept generation*. These three methods are illustrated by using the example of synthesizing the *base concepts*: "tomato" and "snow" in Fig. 5.1, which were used in Chap. 4.

On the other hand, in the field of linguistic studies, it has been revealed that a novel noun–noun phrase can be interpreted through the following three types: ***property mapping***, ***hybrid***, and ***relation linking*** [14]. For example, a knife-fork can be interpreted as follows: a knife-shaped fork, through *property mapping*; one-half as a knife and the other half as a fork, through *hybrid*; and a knife and fork set used together while eating, through *relation linking* (Fig. 5.2).

The three types of linguistic interpretation can be found to correspond to the three methods of *concept synthesis,* as presented in Table 5.1. This correspondence not only makes it possible to compare *concept synthesis* with linguistic interpretation, but also gives validity to the classification of the three methods of *concept synthesis*—particularly the classification between *property mapping* and *concept blending*—by distinguishing the process of transferring a property or feature (*abstract concept*) to the existing concept from the process of blending two *abstract concepts*. In this chapter, in order to compare *concept synthesis* with the linguistic interpretation, the term ***blending*** is used for *hybrid* and *concept blending* and ***thematic relation*** is used for *relation linking* and *concept integration in thematic relation*.

5.2 Concept Synthesis Versus Linguistic Interpretation

We will compare *concept synthesis* with linguistic interpretation in order to identify and capture the essence of the *concept synthesis*. These two processes are completely different; in the former, the generating process is focused on, whereas in the latter, the understanding process is focused on. Although these two

Fig. 5.2 The three types of linguistic interpretations of the noun–noun phrase 'knife-fork'

Table 5.1 Thought types for both linguistic interpretation and concept synthesis

	Property mapping	Blending	Thematic relation
Linguistic interpretation	Property mapping (e.g. 'a knife-shaped fork')	Hybrid (e.g. 'one-half is a knife and the other half is a fork')	Relation linking (e.g. 'a knife and fork set')
Concept synthesis	Property mapping (e.g. "white tomato")	Concept blending (e.g. "powdered ketchup")	Concept integration in thematic relation (e.g. "humidifying refrigerator")

processes involve completely different notions, they can be captured in the same framework. By considering the two *base concepts* in the *concept synthesis* as a compound phrase composed of two nouns (noun–noun phrase), one can compare these two processes. In the field of linguistic studies, many results have been accumulated from the study of noun–noun phrases [1, 3, 14]. Smith et al. [10] reported how people interpreted a noun–noun phrase (e.g. the term 'computer-dog') and addressed the issue that there would be a divergence in interpretations. Further, this term was used as a source of ideas for new inventions by carrying out experiments [12]. The students were required 'to envision what a computer dog would be like, if it were a variant on a mouse' and it was reported that the students produced several novel ideas. We assumed that the methods for the experiment could be developed into a design task for *concept synthesis*.

In order to distinguish the concept expressed by a designer from that imagined in the designer's mind, we introduce the term ***design idea***: an expression with a concrete figure or shape expressed outside the designer's mind.

In our previous trial, in order to compare the *concept synthesis* with linguistic interpretation, it was revealed that the proportion of *property mapping* was lower in design tasks where the subjects were required to generate a new *design idea* from the given *base concepts*, than in interpretation tasks where the subjects were required to interpret the noun–noun phrase which corresponds to the *base concepts*. In contrast, the proportion of *blending* was higher in the design tasks than in the interpretation tasks [11]. This result indicates that *concept synthesis* is characterized by *concept blending*. The reason for this is assumed to be as follows. New concepts generated by *property mapping* are limited from the viewpoint of *innovativeness*, since *property mapping* cannot extend beyond the categories of the given *base concepts*, as mentioned in Chap. 4. In contrast, *concept blending* can generate an *innovative* new concept which extends beyond the existing categories, because the concept generated by this process does not belong to either category of the *base concepts* owing to its definition. Therefore, *concept blending* is assumed to characterize the *concept synthesis* when high originality is pursued. On the other hand, in the linguistic interpretation process, the given phrases are interpreted naturally. Accordingly, it is assumed that *concept blending* is used more in the design task than in the interpretation task.

In order to investigate the essence of *concept synthesis* more closely, we focus on recognition types (**commonality**, **alignable difference**, and **nonalignable difference**). The notions of *alignable difference* and *nonalignable difference* are described as follows [8, 13]. *Alignable differences* are coded for both references to values along a single dimension, such as a sled carries more than one person and a ski carries only one person, as well as for implicit references, such as sleds and skis carry different number of people. *Nonalignable differences* are coded for all other differences that were listed. These differences simply focused on a disparity between the two items without highlighting a common dimension. An example of a *nonalignable difference* would be that an airplane is solid but a puddle is not.

The recognition types are considered to be related to *concept synthesis* as follows. *Property mapping* involves the transfer of some properties or features from an existing concept to the other *base concept*. Therefore, the property or feature focused on in *property mapping* is assumed to be related in an *alignable difference,* since, in *property mapping*, the feature focused on in the existing concept displaces the corresponding feature in the other *base concept*. This displacement implies that both these features involve different values along the same dimension. For example, "white tomato" in Fig. 5.1 is obtained by transferring the feature of "white" to "tomato". Here, the feature "white" is classified as an *alignable difference* since "white" is the value of colour and tomato has another value of colour, which is "red". On the other hand, in *concept blending* and *concept integration in thematic relation*, the features focused on need not be *alignable*. For example, in "powdered ketchup" in Fig. 5.1, the feature "powder" is classified into a *nonalignable difference*, since the corresponding feature of "powder" is thought to be non-recognizable in "tomato". Moreover, in the example of the "humidifying refrigerator" in *concept integration in thematic relation*, "humidifying" is classified into a *nonalignable difference*. Incidentally, it was reported that more *commonalities* and *alignable differences* were listed for similar pairs than for dissimilar pairs, while more *nonalignable differences* were listed for dissimilar pairs than for similar pairs [8, 13].

On the basis of the above considerations, we can assume that the *concept synthesis* and its creativity are characterized by *nonalignable differences* which are analogous to the 'dissimilarity-recognition process'.

5.3 Experiment to Capture the Essence of Concept Generation

5.3.1 Outline of the Experiment

We conducted an experiment to compare *concept synthesis* with the process of linguistic interpretation by focusing on the recognition types (*commonality*, *alignable difference*, and *nonalignable difference*).

In the experiment, the subjects were required to perform the following three tasks: interpret a novel noun–noun phrase (interpretation tasks), design a *new design idea* from the same phrase (design tasks), and list the similarities and dissimilarities between the two nouns (similarity and dissimilarity listing task). The first and second tasks were performed in order to compare the *concept synthesis* with linguistic interpretation. The third task was performed in order to confirm if the recognition types are manifested during the design or interpretation tasks or if they are derived from a subject's trait. Prior to the experiment, we conducted a preliminary experiment to select the noun–noun phrases to be used.

The *design ideas* were analysed according to the thought types (*property mapping*, *blending*, and *thematic relation*), and the **design idea features** (which are provided by the designers who are required to provide a number of words to explain the *design ideas*) were analysed on the basis of the recognition types. Although the *design idea features* are not a *design idea* themselves, we can safely assume that they include the contents of a *design idea*. Further, the creativity in the *design ideas* was examined as follows. First, the *design ideas* were evaluated from the viewpoints of originality and practicality. Second, the *design idea features* and features enumerated by explaining the responses to the interpretation task were judged in order to determine if they were emergent features.

5.3.1.1 Interpretation Tasks

The interpretation tasks consisted of two sub-tasks. First, the subjects were required to naturally interpret the noun–noun phrases (the 'interpretation task'). Second, they were required to enumerate some words (the 'interpretation feature') to explain each interpretation (the 'interpretation feature enumerating task'). The responses to the interpretation task were analysed on the basis of the thought types. The responses to the interpretation feature enumerating task were analysed according to the recognition types and emergence of features.

5.3.1.2 Design Task

The design task also consisted of two sub-tasks. First, the subjects were required to design a new *design idea* from the noun–noun phrases (the 'design task'). They were required to not only draw a sketch but also explain the *design image* in a sentence. In this chapter, hereafter, the term ***design idea*** is used to convey not only the sketch but also the sentence. Second, they were required to enumerate *design idea features* (the 'design feature enumerating task'). The *design ideas* were analysed according to the thought types and creativity (originality and practicality). The *design idea features* were analysed on the basis of the recognition types and emergence of features.

5.3.1.3 Similarity and Dissimilarity Listing Task

In this task, the subjects were required to compare the two nouns in the noun–noun phrase used in the interpretation task (as well as the design task) and to list their common (similarities) and uncommon (dissimilarities) features (the 'similarity and dissimilarity listing task'). Here, common features are those which belong to both the nouns, and uncommon features are those which belong to one noun but not the other. The responses to the similarity and dissimilarity listing task were analysed on the basis of the recognition types.

5.3.2 Method of the Experiment

5.3.2.1 Selecting the Noun–Noun Phrases Used in the Preliminary Experiment

The noun–noun phrases to be used in the preliminary experiment were selected carefully, according to the following procedures.

First, for the 1,055 words listed in the Associative Concept Dictionary [4], the number of *associative concepts* (i.e. concepts associated with the word) was investigated, and the words which had between 168 and 299 (mean ± SD) *associative concepts* were selected in order to control the associative effectiveness (this point will be discussed in Chap. 8); in all, 698 words (i.e. nouns) were selected. Second, these selected nouns were classified into eight categories (furniture, musical instrument, container, natural item, manufactured item, tool, wheeled vehicle, and non-wheeled vehicle) and exceptions were drawn by referring to the method mentioned in Wilkenfeld and Ward [13]. Third, two nouns were randomly combined. Finally, 20 noun–noun phrases were selected. These 20 noun–noun phrases were used in the preliminary experiment.

Table 5.2 Noun–noun phrases used in the interpretation task and similarity and dissimilarity listing task

Noun A	Noun B	Category of noun A	Category of noun B
Ship	Box	Non-wheeled vehicle	Container
Piano	Guitar	Musical instrument	Musical instrument
Desk	Elevator	Furniture	Non-wheeled vehicle
Drawer	Plate	Furniture	Container
Ship	Guitar	Non-wheeled vehicle	Musical instrument
Book	Desk	Manufactured item	Furniture

5.3.2.2 Preliminary Experiment to Select the Noun–Noun Phrases to be Used in the Experiment

In the preliminary experiment, 18 subjects were required to compare the two nouns in a noun–noun phrase and list the common (similarities) and uncommon (dissimilarities) features between the nouns. We planned to select noun–noun phrases such that the number of listed features, both common and uncommon, would be approximately the same and the variance would be large in order to control the effectiveness of the distance between the *base concepts* (this point will be discussed in Chap. 8); this was carried out according to the following guidelines:

- The difference between the mean of the number of common features and that of different features is lower than the average (0.6), which is obtained by calculating the average of the differences between the responses by the subjects.
- The standard deviation of the number of common features is higher than the overall average (1.0), which is obtained by calculating the average of SD.
- The standard deviation of the number of different features is higher than the overall average (1.1), which is obtained by calculating the average of SD.

The following six noun–noun phrases were eventually selected: 'ship-box', 'piano-guitar', 'desk-elevator', 'drawer-plate', 'ship-guitar', and 'book-desk' (Table 5.2). These six noun–noun phrases were used in the interpretation tasks and similarity and dissimilarity listing task.

Next, two noun–noun phrases used for the *base concepts* in the design tasks were selected according to the following guidelines:

- Do not select noun–noun phrases where the same noun is included in the two noun–noun phrases.
- Do not select a noun–noun phrase which can be interpreted as a commonly known phrase.
- Choose a noun–noun phrase which is suitable for the *base concepts* in a design task.

As a result, two noun–noun phrases—'desk-elevator' and 'ship-guitar'—were selected.

5.3.2.3 Subjects

The subjects comprised 22 undergraduate and graduate students who were majors in industrial design. They were equally divided into two groups, Group A and Group B, in order to control the sequence effect of the tasks (interpretation tasks → design tasks; design tasks → interpretation tasks).

5.3.2.4 Procedure of the Experiment

The experiment was conducted using a booklet which included instructions for the interpretation task, interpretation feature enumerating task, design task, design feature enumerating task, and similarity and dissimilarity listing task as well as the answer sheets. Each group was assigned a different room. The experiment was conducted as follows.

- Step 1: Group A performed the interpretation task (1 min for each interpretation; total 6 min), while Group B performed the design task (10 min for each design; total 20 min).
- Step 2: Group A performed the interpretation feature enumerating task (2 min for each interpretation; total 12 min), while Group B performed the design feature enumerating task (2 min for each *design idea*; total 4 min).
- Step 3: Group A performed the design task (10 min for each design; total 20 min), while Group B performed the interpretation task (1 min for each interpretation; total 6 min).
- Step 4: Group A performed the design feature enumerating task (2 min for each *design idea*; total 4 min), while Group B performed the interpretation feature enumerating task (2 min for each interpretation; total 12 min).
- Step 5: Groups A and B performed the similarity and dissimilarity listing task (2 min for each noun–noun phrase; total 12 min).

In the design task, the subjects were required to design a new *design idea* and were informed that the *design ideas* would be evaluated on the basis of originality and practicality. On the other hand, in the interpretation task, they were required to naturally interpret the given phrases.

5.3.3 Method of Analysis

The responses obtained in the experiment were analysed on the basis of the recognition types, thought types, creativity (originality and practicality), and emergence of features. In this experiment, the emergence of the enumerated features was analysed, while the *design ideas* were also evaluated by the raters from the viewpoint of originality and practicality. In order to accurately compare the design task with the interpretation task, only the responses to 'desk-elevator' and 'ship-guitar', which were used in the design task, were analysed.

Table 5.3 Classification standard of recognition types (commonality, alignable difference, and nonalignable difference)

Classification standard and example	
Commonality	When an identified feature refers to the common feature of concept A (or part of concept A) and concept B (or part of concept B) or is associated with both concepts
	Example: In the comparison between "ship" and "guitar", "toy" is judged as a *commonality*, since both "ship" and "guitar" can be toys
Alignable difference	When an identified feature indicates a dimension and the values of each concept are different along the dimension, regardless of it being expressed explicitly or implicitly
	Example: In the comparison between "ship" and "guitar", features related to 'colour' are judged as an *alignable difference*
Nonalignable difference	When an identified feature refers to a feature associated with only one concept (or part of the concept)
	Example: In the comparison between "ship" and "guitar", "vehicle" is judged as a *nonalignable difference*
Other	Cases that do not fall into any of the above-three categories
	Example: In the comparison between "ship" and "guitar", "planter" was judged as a feature that does not fit into any category

5.3.3.1 Classification of Responses According to Recognition Types

We classified the *design idea features*, interpretation features, and responses to the similarity and dissimilarity listing task on the basis of the recognition types for the two nouns in the noun–noun phrases used in each task. This classification was done by the experimenter and a graduate student (who was naïve to the purpose of the experiment) in accordance with the standards set by us in reference to those listed by Markman and Gentner [6, 7]. The classification standard and examples are shown in Table 5.3.

5.3.3.2 Classification of Responses According to Thought Types

The *design ideas* (sketch and sentence) and interpretation were classified by the experimenter and a graduate student (who was naïve to the purpose of the experiment) on the basis of the thought types according to the classification standard presented in Table 5.4; these were formulated and used in our previous trial [11] in accordance with Wisniewski [14]. This classification was used to categorize the *design ideas* and interpretation and was not actually based on the thinking process but on the outcomes. The classification standard is shown in Table 5.4.

5.3.3.3 Creativity Evaluation

The creativity of the *design ideas* (sketch and sentence) was evaluated from the viewpoint of practicality (whether the idea is achievable and feasible) and

Table 5.4 Classification standard of the thought types (property mapping, blending, and thematic relation)

Classification standard and example	
Property mapping	When the response is a 'type of concept B (A) similar to concept A (B)'
	When a response is understood in the framework of 'a part of the property (shape) of concept A (B) or the concept associated with concept A (B) is transferred to concept B (A)'
	Example: In the design task of 'ship-guitar', "ship-shaped guitar" is judged as *property mapping*
Blending	When the response has the properties of both concepts A and B, and it is neither concept A nor concept B
	When the response is related to concept A (B) from the viewpoint of the material or the response is a part of concept A (B), and it has the property of concept B (A)
	Example: In the interpretation task of 'piano-guitar', "thing that is made up of clavier and strings" is judged as *blending*
Thematic relation	When the response stems from a scene in which concepts A and B are related to each other (e.g. A move to B)
	When the response is a 'type of concept B (A) that is made of concept A (B)'
	When the response is a 'type of concept B (A) that is also meaningful with regard to concept A (B)'
	Example: In the design task of 'ship-guitar', "guitar that plays well even on the moving ship" is judged as a *thematic relation*
Other	Cases that do not fall into any of the above-three categories
	Example: In the design task of 'ship-box', "ship" is judged as one that does not fit into any category

originality (whether the idea is novel), referring to the creativity evaluation method given in Finke et al. [2]. In all, 11 raters evaluated all the *design ideas* on the basis of a five-point scale (1: low and 5: high). The rating scores were averaged for each *design idea*. The *design ideas* with lower scores for practicality than the average score for all *design ideas* were excluded from the creativity evaluation. For the remaining *design ideas*, the scores for originality, which were obtained by averaging the scores of all the raters, were considered as the measure of creativity.

5.3.3.4 Judgement of Emergent Features

By referring to Wilkenfeld and Ward [13], the enumerated features (interpretation features and *design idea features*) were judged in terms of whether or not they were emergent features. The Associative Concept Dictionary [4] and synonym dictionary [15] were used for this judgement. When the feature was found to be one which can be associated from either of the two nouns in a noun–noun phrase, it was judged as a non-emergent feature. Furthermore, we investigated the synonyms of the associated concepts by using the synonym dictionary. When the feature was found to be a synonym of the associated concepts of the nouns, it was also judged as a non-emergent feature.

5.3.4 Results

A total of seven responses (three each for the design task and design feature enumerating task and one for the interpretation feature enumerating task) were excluded from the analysis because they were inadequate. We checked the influence of the sequence of the tasks. The proportion of the thought types of Groups A and B did not show a significant difference. For the interpretation and design tasks, the Chi-square values were 0.96, n.s. and 0.24, n.s., respectively.

Examples of the responses are shown in Fig. 5.3.

5.3.4.1 Comparison of the Design and Interpretation
Based on the Thought Types

The classification of the *design ideas* (sketch and sentence) and the interpretation on the basis of the thought types are illustrated in Fig. 5.4. Here, 87.7% of the classifications of the interpretations were agreed on between the two raters, and of the *design ideas*, 90.9%. We found a high proportion of *blending* in the *design ideas* as opposed to the interpretation. This result corresponds to that of our previous trial [11]. The Chi-square test detected a significant difference in the proportion of the thought types between the two tasks ($\chi^2(2) = 9.24$, $p < 0.01$). The result of the

Noun–noun phrase	Interpretation task	Design task
	Response	**Response**
	An elevator to carry a desk, which is placed in a school. A person cannot get on the elevator. This elevator can carry many desks in a smaller space.	A table which can be modified by replacing the surface with the upper and lower levels. Its structure is made up of levels such that each level can be used for dining, operating a computer, or reading a book. This type of table is useful for a person who would not like to use the same table for working on a computer and dining, and he or she does not have enough space for two tables.
Desk-elevator	**Thought type** Thematic relation	**Thought type** Property mapping
	Enumerated Features and Recognition Type	**Enumerated Features and Recognition Type**

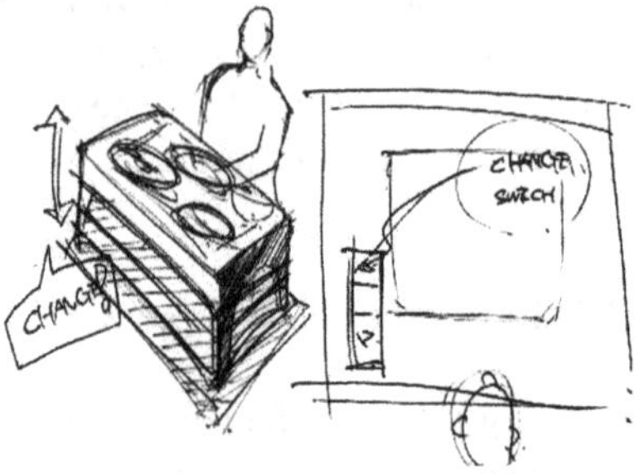

Feature	Recognition Type	Feature	Recognition Type
object	other	button	nonalignable difference
school	commonality	flat	commonality
place	commonality	reading	nonalignable difference
carry	nonalignable difference	change	nonalignable difference
		level	commonality
		switch	nonalignable difference
		dining	nonalignable difference
		up and down	nonalignable difference
		lunch box	nonalignable difference
		interior design	nonalignable difference
		personal computer	nonalignable difference

Fig. 5.3 Examples of the responses

Ship-guitar	**Response**	**Response**
	A guitar of the same scale as that of a ship. It can be used as the emblem of a large town.	
		A guitar using a wave: The string of the guitar is plucked by the driving force of the boat and the waves of the water, thus resulting in a sound.
	Thought type	**Thought type**
	Property mapping	Blending
	Enumerated Features and Recognition Type	**Enumerated Features and Recognition Type**

Feature	Recognition Type		Feature	Recognition Type
fragile	other		leisure	commonality
bright	other		live broadcast	nonalignable difference
large	nonalignable difference		reaction	other
object	commonality		sport	nonalignable difference
coarse	other		exciting	commonality
inspection	commonality		resonance	nonalignable difference
base	other			
long	nonalignable difference			

Fig. 5.3 (continued)

residual analysis indicated a significant difference only in *blending*, as shown in Table 5.5. From this result, it is assumed that *concept blending* characterizes the *concept synthesis* rather than the linguistic interpretation.

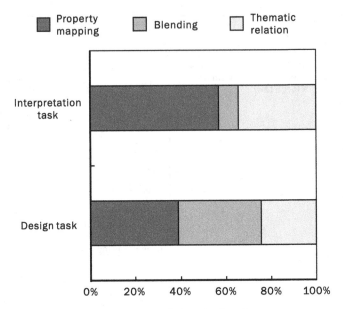

Fig. 5.4 Classification of the responses according to the thought types

5.3.4.2 Comparison of the Design and Interpretation Tasks Based on the Recognition Types (*Commonality, Alignable Difference, and Nonalignable Difference*)

According to the standard presented in Table 5.3, the interpretation features and *design idea features* were classified on the basis of the recognition types. Here, 72.5% of the classifications of the interpretations and the *design idea features* were agreed on between the two raters; further, 80.3% were agreed on for the similarity and dissimilarity listing task, which is described in the next subsection.

The results are illustrated in Fig. 5.5. In the Chi-square test, a significant difference was detected in the proportion of the recognition types between the interpretation features and *design idea features* ($\chi^2(2) = 4.69$, $p < 0.10$). The result of the residual analysis indicated that the proportion of *nonalignable difference* in the *design idea features* was higher than that in the interpretation feature, while the proportion of *commonality* was low (Table 5.6). It is assumed that more attention is paid to *nonalignable difference* in the *concept synthesis* than in linguistic interpretation.

5.3.4.3 Relation between the Thought Types and Recognition Types

First, with respect to the interpretation features and *design idea features*, we determined the proportion of the recognition types for each interpretation and *design idea*. We calculated the average of the proportions of the *design ideas* and

Table 5.5 Result of the residual analysis on the classification of the responses according to the thought types

Thought type	Property mapping	Blending	Thematic relation
Interpretation task	1.64	−3.04 **	0.98
Design task	−1.64	3.04 **	−0.98

| residual | > 1.65 → + $p < 0.10$; | residual | > 1.96 → * $p < 0.05$; | residual | > 2.58 → ** $p < 0.01$

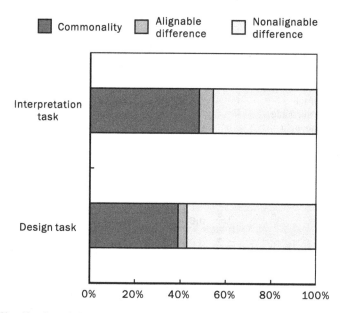

Fig. 5.5 Classification of the responses according to the recognition types

Table 5.6 Result of the residual analysis on the classification of the responses according to the recognition types

Recognition type	Commonality	Alignable difference	Nonalignable difference
Interpretation task	1.79 +	0.85	−2.13 *
Design task	−1.79 +	−0.85	2.13 *

| residual | > 1.65 → + $p < 0.10$; | residual | > 1.96 → * $p < 0.05$; | residual | > 2.58 → ** $p < 0.01$

interpretations classified under each thought type; this result is presented in Table 5.7. A two-factor factorial ANOVA (between-subjects factorial design; factor 1: interpretation or design task, factor 2: thought types) indicated a significant difference only in the factor of the thought type with respect to the proportion of the *nonalignable difference* ($F(2, 76) = 3.22$, $p < 0.05$). This suggests that the thought types may be characterized by *nonalignable difference* (Fig. 5.6).

Table 5.7 Mean of the proportion of recognition types among the responses classified into each thought type for the interpretation task and design task (based on the feature enumerating task)

	Property mapping	Blending	Thematic relation
Commonality			
Interpretation	0.471	0.243	0.497
Design idea	0.448	0.314	0.504
Alignable difference			
Interpretation	0.052	0.125	0.141
Design idea	0.028	0.011	0.053
Nonalignable difference			
Interpretation	0.477	0.632	0.362
Design idea	0.524	0.675	0.443

Fig. 5.6 Mean of the proportion of nonalignable difference in the thought types for the interpretation task and design task (based on the feature enumerating task)

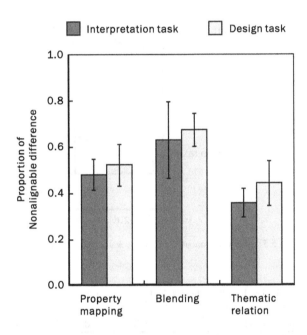

From the results in Figs. 5.4 and 5.5, it was found that *blending* and *nonalignable difference* characterize *concept synthesis*. Confirming this finding, the result obtained above (Fig. 5.6) suggests that *nonalignable difference* is related to *blending*.

Second, with regard to the responses obtained in the similarity and dissimilarity listing task, we determined the proportion of the recognition types for each subject. It is assumed that this proportion indicates the manner in which things or concepts are viewed by the subjects. We calculated the average of the proportions of the responses belonging to each thought type; the thought types were determined on the basis of the corresponding thought types of the *design idea* and interpretation of the subject who responded. This result is presented in Table 5.8. A two-factor

Table 5.8 Mean of the proportion of recognition types among the responses classified into each thought type for the interpretation task and design task (based on the similarity and dissimilarity listing task)

	Property mapping	Blending	Thematic relation
Commonality			
Interpretation	0.392	0.775	0.41
Design idea	0.39	0.451	0.405
Alignable difference			
Interpretation	0.205	0.133	0.186
Design idea	0.249	0.163	0.117
Nonalignable difference			
Interpretation	0.403	0.092	0.405
Design idea	0.361	0.386	0.478

factorial ANOVA revealed that there was no significant difference in the factor of the thought types with respect to the design task.

This result suggests that focusing on *nonalignable difference* is not a trait accumulated in subjects (Fig. 5.7); rather, it occurs with regard to the design and interpretation processes.

5.3.4.4 Comparison of the Design and Interpretation Tasks from the Viewpoint of the Emergence of Features

The mean of the emergent features (interpretation features and *design idea features*), which were judged according to the standard presented in the previous subsection, is illustrated in Fig. 5.8. This figure shows that more emergent features were used to explain the *design idea* rather than to explain the interpretation (two-sided test: $t(82) = 2.36$, $p < 0.05$). This result indicates that more novel features emerge in the *concept synthesis* rather than during the linguistic interpretation process.

5.3.4.5 Relation between Creativity and Recognition Types

The creativity of the *design idea* was evaluated according to the procedure determined in Sect. 5.3.3.3. Kendall's coefficient of concordance showed a significant concordance in both originality and practicality (originality: $W = 0.34$, $\chi^2(40) = 148.86$, $p < 0.01$; practicality: $W = 0.32$, $\chi^2(40) = 142.18$, $p < 0.01$). The number of remaining *design ideas* with higher scores for practicality than the average score of all the *design ideas* is 9 (*property mapping*), 6 (*blending*), and 4 (*thematic relation*).

No correlation between the evaluated score of originality and the proportion of recognition type was detected for all the *design ideas*. However, a marginal

Fig. 5.7 Mean of the proportion of nonalignable difference in the thought types for the interpretation task and design task (based on the similarity and dissimilarity listing task)

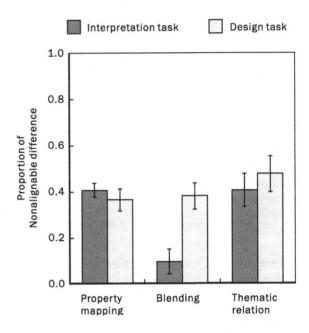

Fig. 5.8 Mean of the number of emergent features

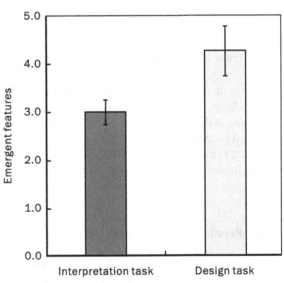

significant correlation was detected between the evaluated score of originality and the proportion of the *commonality* and *nonalignable difference* for the *design ideas* classified into *blending* (*nonalignable difference*: $r = 0.80$, $F(1, 4) = 7.11$, $0.05 < p < 0.10$; *commonality*: $r = -0.80$, $F(1, 4) = 7.11$, $0.05 < p < 0.10$) (Fig. 5.9).

Fig. 5.9 Correlation
between the evaluated score
of originality and proportion
of nonalignable difference in
blending

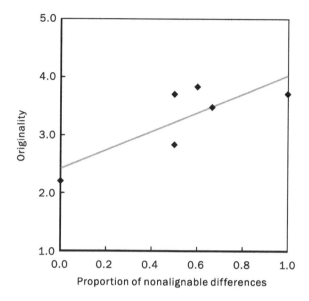

This result indicates that focusing on *nonalignable difference* is related to originality in *concept blending*, which characterizes *concept synthesis*.

5.3.4.6 Relation between Creativity and the Emergence of Features

The relation between the number of emergent features and the evaluated score of originality is illustrated in Fig. 5.10. A regression analysis detected a significant curve regression ($R = 0.68$, $p < 0.01$) rather than a linear regression. This result indicates that there exists an appropriate level of emergent features for high originality in *design ideas*.

5.4 Relevance of the Definition of Concept Generation

This experiment revealed that the main factors in *concept synthesis* are *concept blending* and *nonalignable difference*. This result gives certain relevance to the definition of *concept generation* (*composing* a desirable concept towards the *future*) which is based on the *inner sense–driven phase*—related to *high-order concept generation* which is based on the dissimilarity-recognition process—because *high-order concept generation* involves *concept blending*, and the notion of dissimilarity is analogous to *nonalignable difference*. However, the role of *concept integration in thematic relation,* identified as another important method in *concept generation* (in Chap. 4), was not identified in this experiment, where the

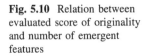

Fig. 5.10 Relation between evaluated score of originality and number of emergent features

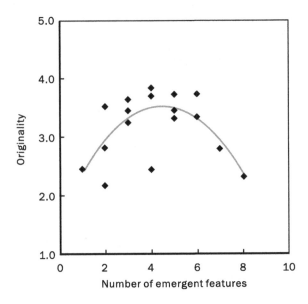

characteristics of the designed outcomes are focused on. The characteristics of *concept integration in thematic relation* are discussed again in Chap. 6, where its thinking process is focused on.

References

1. Costello FJ, Keane MT (2000) Efficient creativity: constraint-guided conceptual combination. Cognit Sci 24:299–349. doi:10.1207/s15516709cog2402_4
2. Finke RA, Ward TB, Smith SM (1992) Creative cognition: theory, research, and applications. The MIT Press, Cambridge
3. Hampton JA (1997) Emergent attributes in combined concepts. In: Ward TB, Smith SM, Vaid J (eds) Creative thought: an investigation of conceptual structures and processes. APA Book, Washington
4. Ishizaki S (2006) Associative concept dictionary (ver 2). Keio University, Fujisawa (CD-ROM, in Japanese)
5. Lubart TI (1994) Creativity. In: Sternberg RJ (ed) Thinking and problem solving. Academic Press, San Diego
6. Markman AB, Gentner D (1993) Structural alignment during similarity comparisons. Cognit Psychol 25:431–467. doi:10.1006/cogp.1993.1011
7. Markman AB, Gentner D (1993) Splitting the differences: a structural alignment view of similarity. J Mem Lang 32:517–535. doi:10.1006/jmla.1993.1027
8. Markman AB, Wisnieski EJ (1997) Similar and different: the differentiation of basic-level categories. J Exp Psychol Learn Mem Cognit 23:54–70. doi:10.1037/0278-7393.23.1.54
9. Rothenberg A (1979) The emerging goddess: the creative process in art, science, and other fields. University of Chicago Press, Chicago
10. Smith SM, Ward TB, Finke RA (eds) (1995) The creative cognition approach. The MIT press, Cambridge

11. Taura T, Nagai Y, Morita J, Takeuchi T (2007) A study on design creative process focused on concept combination types in comparison with linguistic interpretation process. In: Proceedings of the 16th international conference on engineering design. Paris, France, 28–31 August (CD-ROM)
12. Ward TB, Finke RA, Smith SM (2002) Creativity and the mind: discovering the genius within. Basic Books, New York
13. Wilkenfeld MJ, Ward TB (2001) Similarity and emergence in conceptual combination. J Mem Lang 45:21–38. doi:10.1006/jmla.2000.2772
14. Wisniewski EJ (1996) Construal and similarity in conceptual combination. J Mem Lang 35:434–453
15. Yamaguchi T (ed) (2006) Japanese thesaurus dictionary. Taishukan, Tokyo (in Japanese)

Chapter 6
Thinking Pattern in Concept Synthesis (1): Expansion of the Thought Space

Abstract In this chapter, we investigate the thinking pattern in *concept synthesis*; in particular, we focus on the expansion of the thought space. We conduct an experiment using the method of *extended protocol analysis* in order to obtain the in-depth data pertaining to the expressed design activity. The results reveal that there is a marginal significant correlation coefficient between the expansion of the thought space and the evaluated score of originality. Furthermore, we find that the *thematic relations* were used by the designer more frequently in the case of designing a *design idea* with a high-evaluated score of originality than one with a low-evaluated score of originality. From these results, we infer that expanding the thought space on the basis of the association process leads to a highly creative *design idea*; in particular, expanding the thought space by using *thematic relations* is an effective method of creating a highly creative *design idea*.

6.1 Divergent Thinking in Design

'Divergent thinking' has been regarded a basic element in the creative process and is effective in creativity [12]; it is the foundation of the methods not only for evaluation (e.g. creativity test), but also for enhancing creativity at the level of ideation. Joy Paul Guilford, the 3rd President of the American Psychology Society, conducted a psychometric study to understand human creativity. He proposes a theory on the structure of the intellect of an individual and identifies the mental factors of intelligence [4, 5]. From his theory, 'divergent production' is considered the ability to generate multiple solutions for a problem and viewed as the power of creativity. On the basis of the belief of the effectiveness of divergent thinking, techniques which contribute to listing a number of ideas (e.g. Brainstorming by Osborn [10], originally) have been deemed as useful methods for creativity, especially for better ideation, such as to develop original and unique ideas [11].

T. Taura and Y. Nagai, *Concept Generation for Design Creativity*,
DOI: 10.1007/978-1-4471-4081-8_6, © Springer-Verlag London 2013

Divergent thinking is also believed to be useful by many designers [13]. A number of studies on design methods have discussed the roles of divergent thinking on design [6–8]. This tendency is developed because the varieties of thinking directions are believed to lead to novel ideas.

However, a mechanism of divergent thinking in design has not been clarified. Basically, it is difficult to investigate how the thought space expands during *concept generation*. It is necessary to develop a new research method to understand the manner of expansion of the thought space.

6.2 Experiment to Identify the Expansion of Thought Space in Concept Generation

6.2.1 Extended Protocol Analysis Method

In order to investigate the manner of expansion of the thought space, we adopted a new experiment method which was developed by us [14]. This method implements protocol analysis and semi-structured interviews. The Think Aloud Method is usually used for acquiring utterances as protocol data [2]. In the Think Aloud Method, the subjects are required to say aloud what they are thinking while performing a task. The utterances are recorded and the transcript data are analysed. However, in order to understand the manner of the expansion of a designer's thought space, the mechanism of how the designer expands his or her thought space should be clarified. In other words, the reason behind the expressed utterances during the design process should be clarified. It is difficult to obtain data, because the subjects do not always state the reasons behind their activities. In this experiment, the new method of protocol analysis with an explanation of design activities (uttering and drawing), where the subject is required to explain the reason for each design activity while monitoring the video of his or her performance of the design, was adopted.

In addition, the manner of expanding the thought space should be rigorously determined in order to discuss the issue scientifically. We developed a method to project the thought space on Euclidian Space in order to evaluate the degree of expansion of thought space quantitatively.

Furthermore, in order to capture the essence of the expansion of the thought space, the essential nature of the expanding process should be examined. In this experiment, we analysed the expanding process from the viewpoint of the association process of concepts, particularly by focusing on the relation type between concepts. As we pointed out in Chap. 4, the *thematic relation* as well as taxonomical relation is expected to lead to a highly creative idea.

6.2.2 Procedure of the Experiment

We conducted an experiment to clarify the following two issues.

- A quantitative examination of the relation between the expansion of the thought space and creativity of the designed outcome.
- An examination of whether the association process to expand the thought space is based on *thematic relations* or taxonomical relations.

The experiment consisted of two sessions: the design session and interview session.

- Design session (10 min)

The subjects were required to perform the design tasks of *concept synthesis* in the Think Aloud Method, and the utterances and the sketch were recorded using a tape recorder and video camera, respectively. The *base concepts* for *concept synthesis* were selected by referring to the research of Wisniewski and Bassok [16].

- Task A: Design a new furniture piece from "cat" and "hamster"
- Task B: Design a new furniture piece from "cat" and "fish"

- Interview Session (approximately 30 min)

The subjects were required to explain the reason for each design activity (uttering and drawing) while monitoring the video of the design session. The purpose of this session is to determine the reasons why the subjects performed each design activity during the design task (Questions included 'Why did you utter this word?', 'Why did you draw this line?', and so on).

- Creativity evaluation

The creativity of the designed outcomes was evaluated from the viewpoints of practicality (whether the idea is achievable and feasible) and originality (whether the idea is novel) on a five-point scale, referring to the creative evaluation method given in Finke et al. [3]. Only the designed outcomes with practicality scores higher than 3.0 were evaluated from the viewpoint of originality.

6.2.3 Results of the Experiment

The experiment was conducted with three subjects: two were graduate students majoring in industrial design or another discipline and one was a professional artist not majoring in industrial design. In total, fifteen *design ideas* were obtained. Because some subjects were not majoring in industrial design, creativity was evaluated on the basis of the content of the *design ideas* in order to avoid the influence of the drawing skill. The experimenters prepared summaries (written in sentences; in this chapter, these sentences are called ***design concepts***) of the *design ideas*. As examples, three *design concepts* of each task (No. 3, 6, 8, 9, 11, and 15) are presented below.

Task A: Design a new furniture piece from "cat" and "hamster"

• *Design concept* No. 3

"Travelling bag that cares for the pet during travel"
A panel is attached on the side, and an image of the pet is displayed when the panel is opened. Some buttons on the panel enable food to be given to the pet or the pet to be fondled.

• *Design concept* No. 6

"Chair that can be folded like an umbrella"
A chair that can be folded using the mechanism of a folding umbrella and that can be stored in a narrow space. It is possible to store it in an underground compartment after use.

• *Design concept* No. 8

"Revolving shoebox"
This is a doughnut-shaped and life-size rotary shoebox. It rotates when the user stands in front of it, and shoes can be chosen. It is easy to pick the shoes appropriate for an outfit because the bottom of the shoebox is transparent.

Task B: Design a new furniture piece from "cat" and "fish"

• *Design concept* No. 9

"Sideboard with a monitor"
Usually an image of fish in an aquarium is displayed on the monitor. However, it is also a television that can be operated using a remote control. The monitor is at eye level when the viewer is sitting on a chair.

• *Design concept* No. 11

"Fish-tank with casters"
There are legs like those of a chair attached to the bottom of the fish-tank. Since they have casters, it is possible to move the tank easily.

• *Design concept* No. 15

"Sea cushion"
This cushion can float in the sea. It is possible to sit and sleep on it. Moreover, many cushions can be joined together to form a lounger.
Figure 6.1 is an illustration of the sketch of No. 15.

6.2.4 Creativity Evaluation of Design Concepts

The *design concepts* were evaluated by 8 raters (4 of whom were experienced in design). The rating scores were averaged for each *design concept*. Kendall's

Fig. 6.1 Sketch of design idea No.15 "sea cushion"

coefficient of concordance showed a significant concordance in both originality and practicality (originality: $W = 0.28$, $\chi^2(14) = 30.87$, $p < 0.01$; practicality: $W = 0.48$, $\chi^2(14) = 54.22$, $p < 0.01$). The practicality scores of the *design concepts* of No. 1, No. 2, No. 4, No. 7, No. 12, and No. 13 were less than 3.0, whereas the remaining nine satisfied the required practicality score. For these nine *design concepts,* the scores for originality, which were obtained by averaging the scores of all the raters, were considered to be the measure of creativity.

- No. 3: "Travelling bag that cares for the pet during travel"
- No. 5: "Desk-chair"
- No. 6: "Chair that can be folded like an umbrella"
- No. 8: "Revolving shoebox"
- No. 9: "Sideboard with a monitor"
- No. 10: "Case for marine sports"
- No. 11: "Fish-tank with casters"
- No. 14: "Water-tank table"
- No. 15: "Sea cushion"

Table 6.1 presents the evaluated score for creativity of these nine *design concepts.*

6.2.5 Expansion of Thought Space and Creativity

To examine the expansion of the thought space, the nouns (which were determined as newly uttered nouns) were extracted from the utterances.

Next, we determined how the thought space quantitatively expanded during the *concept synthesis.* We calculated the degree of expansion of the thought space, by focusing on the distance between nouns: the distance between newly uttered nouns and the *base concepts* ("cat" and "hamster", "cat" and "fish") and the distance between the newly uttered nouns and "furniture". The distance was measured using the concept dictionary EDR Concept Dictionary [1]. In this dictionary,

Table 6.1 Evaluated creativity scores of nine selected design concepts

No.	Task	Practicality	Originality	Order of high creativity
3	A	3.8	2.9	6
5	A	3.0	2.4	8
6	A	4.1	3.9	1
8	A	3.0	3.6	2
9	B	4.3	2.6	7
10	B	3.8	3.5	3
11	B	4.1	2.0	9
14	B	4.3	3.0	4
15	B	4.1	3.0	5

words are structured in a hierarchical manner. The distance between two words was measured by counting the least number of steps to arrive at one noun from the other noun (Fig. 6.2). The distance from each newly uttered noun to the *base concepts* was determined by adopting the shorter of the two distances to the *base concepts*. For instance, the distance between 'vehicle' and 'blanket' is counted as seven steps (four steps from 'vehicle' to 'inanimate object' and then down three steps to 'blanket'; Fig. 6.2).

By plotting the distance between the newly uttered noun and the *base concepts* in the abscissa axis and the distance between the newly uttered noun and "furniture" in the ordinate axis, the thought space can be scattered in the 2D space. The scatter charts for the *design concepts* of No. 6 and No. 11 are shown in Figs. 6.3 and 6.4, respectively. No. 6 is the *design concept* with the highest evaluated score of originality, and No. 11 is the *design concept* with the lowest evaluated score of originality.

From these figures, it is understood that the thought space of the *design concept* with a high-evaluated score of originality expands more from the origin than the thought space of the *design concept* with a low-evaluated score of originality. In order to discuss the manner of expansion of the thought space quantitatively, we defined the degree of the expansion of thought space as follows.

$$\sum_{i=1}^{N} \frac{\sqrt{x_i^2 + y_i^2}}{N} \tag{6.1}$$

(N is the number of newly uttered nouns. x_i and y_i indicate the co-ordinates on the abscissa axis and ordinate axis, respectively).

Figure 6.5 shows the scatter chart of the calculated expansion of the thought space with the evaluated score of originality. In this relation, a marginal significant correlation coefficient ($r = 0.622$, $F(1, 7) = 4.57$, $0.05 < p < 0.10$) was found. This result indicates that expanding the thought space leads to a highly creative *design idea*.

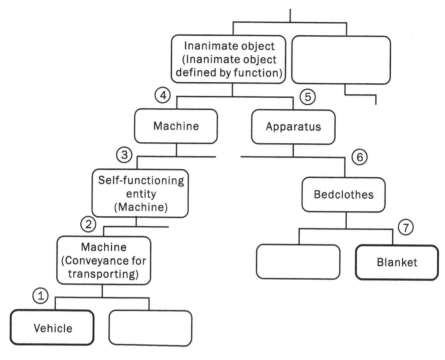

Fig. 6.2 Distance between the nouns in the concept dictionary (in this case, the distance between 'vehicle' and 'blanket' is determined as 7)

Fig. 6.3 Thought space for design concept No. 6

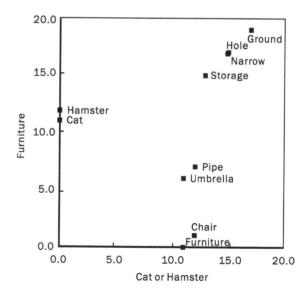

Fig. 6.4 Thought space for design concept No. 11

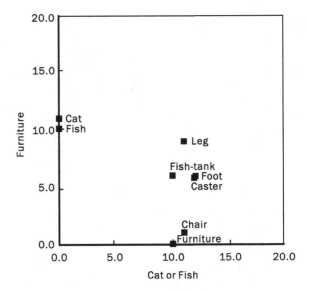

Fig. 6.5 Correlation between the evaluated originality score and the degree of expansion of the thought space

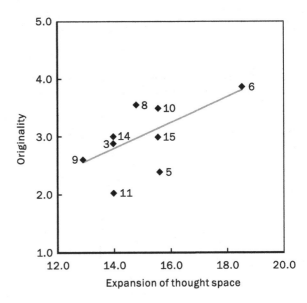

6.2.6 Association Process to Expand the Thought Space

To capture the essence of an expanding thought space, we analysed the types of relations between uttered nouns by focusing on the types of relations (*thematic relation* or taxonomical relation). As pointed out in Chap. 4, *thematic relation* is expected to play an important role in creative *concept generation*.

Table 6.2 Example of the types of relation between two sequential uttered nouns for No. 6

Sequential nouns	Type of relations	Subjects' explanation
Pipe—umbrella	Thematic	Structure of an umbrella
Umbrella—folding umbrella	Taxonomical	A kind of umbrella
Hole—ground	Thematic	A hole in the ground
Ground—narrow space	Thematic	Digging a small hole in the ground
Narrow space—umbrella	Thematic	An umbrella which goes into a gap
Chair—umbrella	–	A chair is an umbrella
Folding umbrella—ground	Thematic	Pulling out a folding umbrella from the ground

Table 6.3 Comparison of the association process between design ideas with high and low originality

	No. 6	No. 11
Evaluated originality score	3.9 (highest)	2.0 (lowest)
Degree of expansion of thought space	18.5 (highest)	13.9 (lowest)
Number of uttered nouns	42	37
Number of thematic relations between sequential uttered nouns	13 pairs (31.0%)	6 pairs (16.2%)

We identified whether each relation between one uttered noun and the next noun belonged to *thematic relation* or taxonomical relation for *design concept* No. 6, with the highest evaluated score of originality, and *design concept* No. 11, with the lowest evaluated score of originality. These relations were identified by referring to the explanation obtained in the interview session: whether the subject explained the relation of the concepts from the viewpoint of a scene, conditions, or functions as the reasons behind the design activity. An example of the identified types of relations is shown in Table 6.2. The result is summarized in Table 6.3. This result shows that *thematic relations* are used by the designer more frequently in the case of designing a *design idea* with a high-evaluated score of originality than in the case of designing a *design idea* with a low-evaluated score of originality.

From the results, it is inferred that expanding the thought space on the basis of the association process leads to a highly creative *design idea*; in particular, to expand the thought space by using *thematic relations* is an effective element in the obtaining of a highly creative *design idea*.

6.3 Paradox of Divergent Thinking

Although, divergent thinking has been thought to be a basic element of the creative process and effective in creativity, as stated in the beginning of this chapter, there is another nearly completely opposing discourse. Weisberg [15] doubts the

(a)

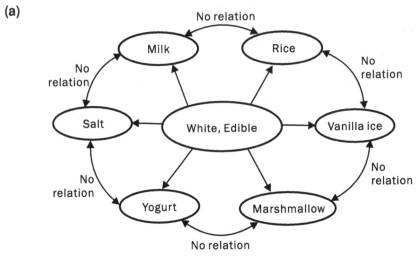

(b)

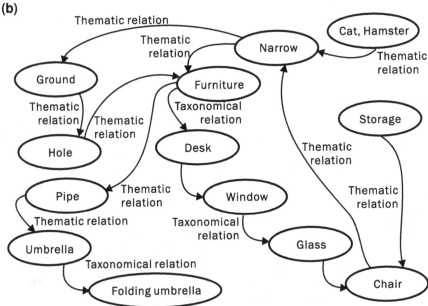

Fig. 6.6 Differences between the divergent thinking test (**a**) and expansion of the thought space in concept synthesis (**b**)

effectiveness of divergent thinking in highly creative people: scientists, artists, architects, and so on. He points out that the score of the test on divergent thinking cannot identify the achievements of such professionals with regard to their creativity, according to the psychological survey developed by Mansfield and Busse [9].

This is the paradox of divergent thinking since a highly creative idea is definitely expected to be obtained from the variety of ideas conceived, as addressed in Chap. 4. Is divergent thinking effective in developing a creative idea or not?

We think the key to solving this paradox exists in the manner of how the thought space expands. In the creativity evaluation test (AC Test of Creative Ability), which is the basis of divergent thinking, responses might be listed without any relation between them. Quick response without any context is expected. On the other hand, in our experiment, the thought space is assumed to expand by relating one concept to another in the association process. In particular, *thematic relations* were found to play an important role in this association process. In other words, the expansion of the thought space has occurred during the search for various ideas by relating different concepts.

Whether the thinking process 'relates' one concept to another or not is thought to be an essential factor in creative *concept synthesis* (Fig. 6.6). The key to solving the paradox of divergent thinking is believed to exist in the absence of 'relation' in the process of the creativity evaluation test.

References

1. EDR Concept Dictionary (2005) EDR electronic dictionary. National Institute of Information and Communications Technology, CPD-V030 (CD-ROM)
2. Ericsson KA, Simon HA (1984) Protocol analysis. The MIT Press, Cambridge
3. Finke RA, Ward TB, Smith SM (1992) Creative cognition: theory, research, and applications. The MIT Press, Cambridge
4. Guilford JP (1956) A factor-analytic study of verbal fluency: studies of aptitudes of high-level personnel. University of Southern California, Columbia
5. Guilford JP (1967) The nature of human intelligence. McGraw-Hill Education, New York
6. Howard T, Dekoninck EA, Culley J (2010) The use of creative stimuli at early stages of industrial product innovation. Res Eng Des 21:263–274
7. Jones JC (1970) Design methods: seeds of human futures. Wiley, New York
8. Maffin D (1998) Engineering design models: context, theory and practice. J Eng Des 9:315–327. doi:10.1080/095448298261462
9. Mansfield RS, Busse TV (1981) The psychology of creativity and discovery: scientists and their work. Nelson-Hall, Chicago
10. Osborn AF (1957) Applied imagination: principles and procedures of creative problem solving. Charles Scribner's Sons, New York
11. Parnes SJ, Meadow A (1959) Effects of 'brainstorming' instructions on creative problem solving by trained and untrained subjects. J Edu Psychol 50:171–176. doi:10.1037/h0047223
12. Runco MA (1991) Divergent thinking. Ablex Publishing, New York
13. Sutton RI, Hargadon A (1996) Brainstorming groups in context: effectiveness in a product de-sign firm. Admin Sci Q 41:685–718
14. Taura T, Yoshimi T, Ikai T (2002) Study of gazing points in design situation—a proposal and practice of an analytical method based on the explanation of design activities. Des Stud 23:165–186. doi:10.1016/S0142-694X(01)00018-7
15. Weisberg RW (1986) Creativity: genius and other myths. WH Freeman and Co, New York
16. Wisniewski EJ, Bassok M (1999) What makes a man similar to a tie? stimulus compatibility with comparison and integration. Cognit Psychol 1:208–238

Chapter 7
Thinking Pattern in Concept Synthesis (2): Complexity of the Thinking Process

Abstract In this chapter, we conduct a computer simulation in order to capture the characteristics or patterns in *concept synthesis*, which may lead to a creative *design idea*. This approach employs a research framework called *constructive simulation*, which may be effective in investigating the generation of a concept—a process which is difficult to observe externally or internally. In the simulation, first, the virtual concept synthesis process is constructed on a semantic network by tracing the relationships between its governing concepts. Next, the relevance of the constructed process is confirmed by using its network structure. The statistical results indicate that the thinking process in which both explicit and 'inexplicit' concepts are 'intricately intertwined' may lead to a creative *design idea*.

7.1 Constructive Simulation Versus Imitative Simulation

Many studies have focused on the characteristics or patterns of the thinking process in design, and various factors which affect this process have been deduced from these studies. Goldschmidt [15] introduces the idea of 'linkography', a method which visualizes a designer's thought characteristics or patterns by forming links between the small units constituting the design protocol. Many studies have used linkography to identify the factors affecting the thinking process in design [17, 29]. Kokotovich [18] proposes another method—nonhierarchical mind mapping—for visualizing a designer's thinking patterns, and reports that it has helped guide novice designers adopt the design thinking framework of expert designers.

From a fundamental viewpoint, it is difficult for people to observe their internal thinking processes by themselves while being deeply engaged in their work because they are then fully engrossed in their work: a mental state described as 'flow' [8], as we addressed in Chap. 3. On the other hand, external observation of the thinking process in design may fail to grasp the inner state of a designer's mind while he or she is deeply engaged in the work, because such a state is stimulated by intrinsic motivation [1, 19] and is formed by invisible inner dynamics [30]. In other words, the realm of a designer's thinking is formulated internally [21]. For these reasons, it is difficult to observe the thinking process in design objectively, whether externally, or internally.

Thus far, the methods used for investigating the characteristics or patterns in this process have generally been based on 'protocol analysis' [10], as mentioned in Chap. 6. Although protocol analysis is an effective means of understanding a designer's thinking process, it is a phenomenological analysis; hence, it occasionally presents difficulties in analysing the design thinking process; particularly, the inner state of the designer's mind that is formed implicitly or 'inexplicitly' (this point will be discussed in Sect. 7.3.5) while people are deeply engaged in their work. As an attempt to overcome this difficulty, the *extended protocol analysis method* was introduced in Chap. 6.

Another approach to such a difficult-to-observe creative thinking process in design is to employ a computational method. A computational method is also effective for understanding the creative thinking process in design in a reproducible manner. In addition, the process can be supported by a computer. Although computational methods have been developed (e.g. [4, 7, 9, 25]), further attempts in this regard will aid in constructing a computational simulation which can capture the characteristics or patterns in the process of *concept generation* which lead to a highly creative *design idea*.

In this chapter, we conduct a computer simulation of the creative process of *concept synthesis* on the basis of a research framework called *constructive simulation*, developed by us [28]. *Constructive simulation* is a framework which could be considered the converse of the more widely known 'imitative simulation' which attempts to model a targeted phenomenon on a computer as accurately as possible, so that the future behaviour of the phenomenon can be predicted. Such computer simulation has already been applied to numerous phenomena, and its accuracy has been verified by comparing the behaviour of the simulation to that of a real-life phenomenon. On the other hand, in *constructive simulation*, the difficult-to-observe mechanism which forms an observable real-life phenomenon is inferred from a different mechanism. When a phenomenon which is similar or analogous to the real-life phenomenon emerges from a different mechanism in the simulation, we believe this indicates that the two mechanisms share certain essential features. If we develop a computer program (mechanism) from which emerges a phenomenon similar or analogous to that of the actual design thinking process or outcome of the design, and if we infer the characteristics or patterns in the actual design thinking process from the characteristics of the computer program, then this method is a *constructive simulation*. The mechanism developed in such a *constructive simulation* is referred to as a 'constructive mechanism'.

The schematics in Fig. 7.1 provide a visual comparison of imitative and *constructive simulation*.

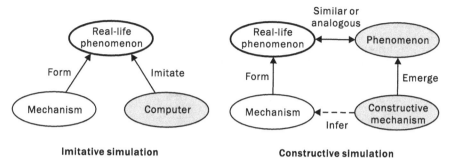

Fig. 7.1 Imitative simulation and constructive simulation

7.2 Method for Constructive Simulation of Concept Synthesis

We seek to investigate the characteristics or patterns in the creative process of *concept synthesis* using computer simulation. However, it is not easy to deductively construct a model for computer simulation by using known knowledge, because the creative process of *concept synthesis* targeted by the simulation is difficult to observe externally or internally. Therefore, we conduct the simulation within the framework of *constructive simulation*, which proceeds according to the following phases:

- Phase 1: Construct a virtual modelling of the *concept synthesis* process on a computer (hereafter, referred to as 'virtual concept synthesis process') as a constructive mechanism.
- Phase 2: Confirm the relevance of the virtual concept synthesis process.
- Phase 3: Verify the validity of the virtual concept synthesis process.
- Phase 4: Infer the characteristics or patterns in the actual concept synthesis process from the characteristics of the virtual concept synthesis process.

The framework for *constructive simulation* of the *concept synthesis* is illustrated in Fig. 7.2, which schematically presents how the four phases mentioned above are coordinated. The procedure for Phases 2, 3, and 4 is explained in the following sections. With regard to Phase 4, the inference is conducted according to the following steps: first, the characteristics or patterns in the virtual concept synthesis process are listed, and then, the characteristics or patterns in the actual concept synthesis process are inferred from the listed characteristics of the virtual concept synthesis process by referring to previous studies on related topics.

7.2.1 Target Concept Synthesis Process

We 'virtually' connect the paths between the starting points (the *base concepts* expressed in words) and the ending points. In order to develop this model, we use

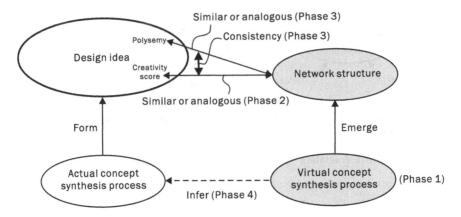

Fig. 7.2 Framework of a constructive simulation of the concept synthesis process

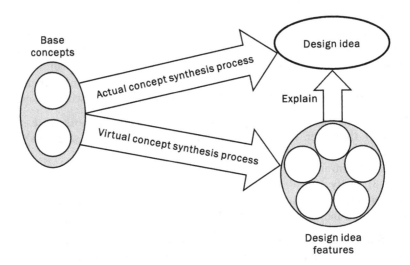

Fig. 7.3 The virtual concept synthesis process

sets of words (*design idea features*) for the ending points, as illustrated in Fig. 7.3. The *design idea features* are provided by designers, who were required to provide a number of words to explain their *design ideas*. Although the *design idea features* are not the *design idea* itself, we can safely assume that they include the contents of the *design idea*.

These sets of connected paths establish the 'virtual concept synthesis process', which plays the role of a constructive mechanism in the framework of *constructive simulation*. The concepts found on each of these paths, apart from the *base concepts* and *design idea features*, are called 'virtual concepts'. These virtual concepts, which are searched for using the steps given below, are exclusively 'virtual'. The main issue in this simulation is the establishment of a connection between the 'virtually' constructed process and the real-life design phenomenon.

7.2.2 *Method of Constructing the Virtual Concept Synthesis Process*

We use a semantic network in order to connect the paths between the words of *base concepts* and the sets of words of *design idea features*. Thus, the virtual concept synthesis process is represented as a network comprising nodes (words) which are found on the paths and links. Many researchers have employed such a notion of 'network' in a variety of applications including the analysis of human language structure [4], semantic network structures [27], and cognitive insight models [25]. Several recent studies in the field of design have also utilized semantic networks [5, 14]. Semantic networks comprise the semantic relationships between words, such as superordinate–subordinate relations and other such associations. In actual practice, semantic networks have proven useful when searching for links between words. Hence, we use a semantic network in order to search for virtual chain processes of words, from the words of *base concepts* to the sets of words of *design idea features*. In other words, we seek to represent the virtual concept synthesis process as a part of a semantic network.

In this simulation, WordNet is used as a semantic network. WordNet [12] is a massive English electronic lexical database which contains information on the manner in which humans process language and concepts. Currently, the database comprises over 150,000 words which are hierarchically organized and interconnected on the basis of various semantic relationships. In WordNet, words are classified into groups of synonyms known as 'synsets'. The words are defined in a simple manner, and the relationships between words are hierarchically described. These relationships include superordinate–subordinate relations and part-whole (meronymic) relations. Because the semantic network used in this simulation must contain a sufficient number of well-organized words, we chose to use WordNet, because it comprises a sufficient number of well-organized words which can simulate the virtual concept synthesis process.

Figure 7.4 schematically presents the flow of steps involved in modelling the virtual concept synthesis process. The modelling method depicted above comprises the following three steps in which each *base concept* and *design idea feature* is expressed as a word. b_i represents the *base concept*; B, a set of *base concepts*; s, a designer; f_j, a *design idea feature*; and $F(s, B)$, a set of *design idea features*. Further, when k is an identifier of meaning, $w{:}k$ indicates a meaning of the word w, and $M(w)$ is a set of meanings.

- Step 1: Search for a short path between a *base concept* and *design idea feature* in a semantic network (the upper diagram box in Fig. 7.4). Figure 7.5 illustrates a schematic of this process of searching for a path on WordNet. In a semantic network, each node represents a meaning of a given word, and each link represents a semantic relationship between the upper and lower layers. By using a semantic network, we can extract the routes between *base concepts* (b_i) and *design idea features* (f_j) by extracting the routes (*Paths*) between the meanings in $M(b_i)$ and $M(f_j)$. If two or more equally short paths are detected, we extract the route which is linked to the meaning with a higher frequency of use in WordNet; the meanings of each word are placed in order of frequency of use in WordNet. Here, when v_1 is the starting node in the path and n is the number of nodes, the path is represented as $Path(v_1, v_n)$. The set of pairs of nodes $Path(b_i : k, f_j : h)$, where $b_i : k \in M(b_i)$, $f_j : h \in M(f_j)$, is represented as

$$Path(b_i : k, f_j : h) = \{(b_i : k, v_2), (v_2, v_3), \cdots, (v_{n-1}, f_j : h)\}. \qquad (7.1)$$

Similarly, we can find the *Path* between any two meanings in $M(b_i)$ and $M(f_j)$ of b_i and f_j. We can then extract the *Paths* for all combinations of the *base concepts* in B and the design features in F.

- Step 2: The node is replaced by a word with the most general meaning and is then extracted as c_l^m. Here, m is an identifier of *Path*, and l denotes the order of the node which appears in the path. A set of paths ($Path(b_i, f_j)$) from b_i to f_j is described as

$$Path(b_i, f_j) = \{\{(b_i, c_2^0), (c_2^0, c_3^0), \cdots, (c_{n-1}^0, f_j)\}, \{(b_i, c_2^1), (c_2^1, c_3^1), \cdots,$$
$$(c_{n-1}^1, f_j)\}, \cdots \cdots\} \qquad (7.2)$$

- Step 3: The virtual concept synthesis process is constructed as $Path(B, F) = \cup_{i,j} Path(b_i, f_j)$, which is a sum of the sets of the paths obtained in Step 2.

7.2.3 Method of Confirming the Relevance of the Virtual Concept Synthesis Process

In the framework of a *constructive simulation*, the virtual concept synthesis process is considered 'relevant' if the phenomenon which emerges from it is found to be 'similar or analogous' to the real-life phenomenon. We use the statistical criteria of the network structure as a phenomenon which has emerged from the virtual concept synthesis process, since the virtual concept synthesis process is constructed on a semantic network and is thus expected to have a close affinity to network theory. Within the wide range of statistical criteria applied in network theory, we chose to employ the statistical criteria of the network structure used by Steyvers and Tenenbaum [27] because the criteria used in their study were selected in order to investigate the structure of semantic networks and determine whether they are inevitably different from other complex natural networks. Their results

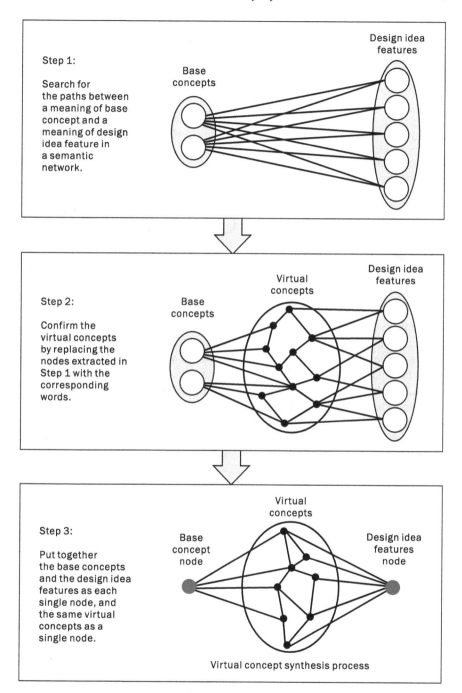

Fig. 7.4 Flow of a modelling of the virtual concept synthesis process

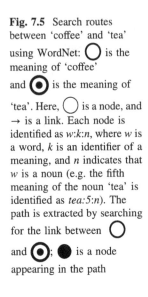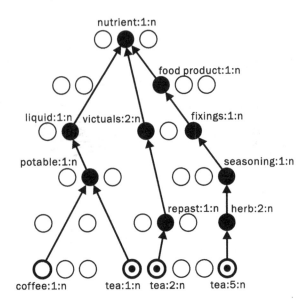

Fig. 7.5 Search routes between 'coffee' and 'tea' using WordNet: ◯ is the meaning of 'coffee' and ◉ is the meaning of 'tea'. Here, ◯ is a node, and → is a link. Each node is identified as *w:k:n*, where *w* is a word, *k* is an identifier of a meaning, and *n* indicates that *w* is a noun (e.g. the fifth meaning of the noun 'tea' is identified as *tea:5:n*). The path is extracted by searching for the link between ◯ and ◉; ● is a node appearing in the path

indicate that semantic networks resemble other complex natural networks in terms of structure. Because we use a semantic network for the virtual modelling of the *concept synthesis*, we can consider the constructed virtual concept synthesis process to be a part of the semantic network. In addition, the actual design thinking process, which is a target of the virtual concept synthesis process, is also part of the designer's thinking, which is in turn a part of a complex natural network. Therefore, by applying the criteria used by Steyvers and Tenenbaum [27], we expect that the characteristics of the virtual concept synthesis process can be determined.

Accordingly, we apply the following criteria to analyse the structural characteristics of the virtual concept synthesis process.

- *n*, *<k>*, and *Density*: These criteria indicate the expansion of the virtual concept synthesis process.
- *C*, *L*, and *D*: These criteria indicate the complexity of the virtual concept synthesis process.

It has been found that expansions of the designer's thought space are related to evaluated creativity scores for *design ideas*, as mentioned in Chap. 6. Therefore, the degree of the expansion of the virtual concept synthesis process can contribute to the production of creative *design ideas*. Furthermore, human knowledge is considered a complex network comprising relationships between various types of information, and such a complex body of information is assumed to manifest in the design thinking process. Therefore, it is assumed that the level of complexity of a network in the virtual concept synthesis process could similarly lead to creative *design ideas*.

Table 7.1 Definitions of criteria used in the network theory

Criteria	Definition
n	Number of nodes
$<k>$	Average degree (degree = number of links)
C	Clustering coefficient
L	Average length of shortest path between a pair of nodes
D	Diameter of network (the longest path among all shortest paths)
Density	Sparseness of network (the situation in which a node is connected to other nodes)

On the basis of the above considerations, we assume that the degree of expansion and complexity of the virtual concept synthesis process have a positive correlation with the evaluated creativity score of *design ideas*.

Table 7.1 summarizes the definitions of criteria used in this simulation. Each criterion is briefly explained in the following manner.

The number of nodes n denotes the number of concepts expressed as words which appear in each virtual concept synthesis process.

The number of links joined to a node is referred to as 'degree', and the average degree $<k>$ is the average number of links joining a node in the network, that is, the value of the total degree over the number of nodes n. A network comprising numerous nodes and links is considered large and assumed to have the capacity for expansion. Thus, these criteria indicate the degree of expansion of a network.

Further, two joined nodes are called 'neighbours'. The probability that the neighbours of an arbitrary node are also neighbours of each other is estimated on the basis of the clustering coefficient C. In terms of network topology, a high probability indicates that there are 'shortcuts' or 'triangles' in the network, a phenomenon which is common in complex networks. In other words, C indicates the complexity of the network. We calculate C from the average C_i in the following manner:

$$C_i = T_i \left/ \binom{k_i}{2} \right. = 2T_i/k_i(k_i - 1), \tag{7.3}$$

where T_i denotes the number of links between the neighbours of node i; k_i, the number of nodes connected to node i; and $k_i(k_i - 1)/2$, the number of links expected between the neighbours of node i if they form a fully connected subgraph. Figure 7.6 illustrates the calculation of C. In this example, the number of nodes (neighbours) connected to Node 1—that is, k_1—is 3, and the number of links which would be expected between neighbours is $k_1(k_1 - 1)/2 = 3(3 - 1)/2 = 3$. If only two neighbours are connected, $T_i = 1$, then, $C = 1/3$.

L denotes the average of the shortest (or geodesic) paths between nodes within the entire network. Conversely, the diameter of the network, D, denotes the longest path within the set of shortest paths. L and D are assumed to be inversely proportional to C. Figure 7.7 schematically presents the relationship between L (or D) and C. In Fig. 7.7a, the length of the shortest path from node 1 to nodes 2, 3, and 4 is one, two, and three, respectively. Then, L is 1.7 and D is 3. In this case, C is zero. In Fig. 7.7b,

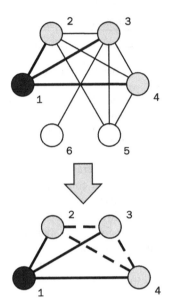

Number of links among neighbours of
Node 1 and the clustering coefficient

$$
\begin{aligned}
0 &\Rightarrow C_1 = 0/3 = 0 \\
1 &\Rightarrow C_1 = 1/3 = 0.3 \\
2 &\Rightarrow C_1 = 2/3 = 0.7 \\
3 &\Rightarrow C_1 = 3/3 = 1.0
\end{aligned}
$$

Clustering coefficient of each node and the average

Node, i	1	2	3	4	5	6	Average
C_i	1.0	0.8	0.5	0.8	1.0	0	0.7

Fig. 7.6 Example of the calculation of C

the length of each of the shortest paths between any two nodes is one. Therefore, the value of both L and D is 1. In this case, the nodes are fully connected; that is, the clustering coefficient is one. Thus, C would be affected by L (or D). Therefore, either L or D can also indicate the degree of complexity of the network.

The *Density* of a network indicates the sparseness of the links in the network. It is obtained by dividing $<k>$ by the size n of the network. Thus, when the relationships (links) between nodes in a network are sparse, the network's *Density* is low.

Summarizing the above consideration, n, $<k>$, and C are expected to have a positive correlation with the evaluated creativity score for *design ideas*, while L, D, and *Density* are expected to have a negative correlation.

Fig. 7.7 Comparison of
networks with long (**a**) and
short lengths (**b**) in the
shortest path

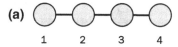

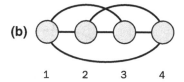

7.2.4 Method of Synthetic Verification of the Virtual Concept Synthesis Process

Because the target mechanism (actual concept synthesis process) is difficult to observe in the first place, it is also difficult to verify the validity of the virtual concept synthesis process. Despite this difficulty, we have attempted to verify the validity of this process using the following two methods. In the first method, we confirm the relevance of the virtual concept synthesis process with regard to the second real-life phenomenon and the consistency between the two relevant situations. In order to do this, we employed the notion of 'polysemy' (multiple meanings of a word) as the second real-life phenomenon, since polysemy is believed to represent another important element of creativity. In general, creativity is assumed to be related to the richness of the related polysemy [11]. Furthermore, it has been noted that the association process between one concept and another forms the basis of the design thinking process, as mentioned in Chap. 6, and it has been assumed that the complex process of association leads to the richness of the polysemy of a *design idea.*

Accordingly, we examine the relationship of the polysemy (number of meanings) of the *design idea features* to the structural criteria of the virtual concept synthesis process (the second relevant situation) and to the evaluated creativity score of the *design ideas.* We do this in order to confirm the relevance of the virtual concept synthesis process with regard to the second real-life phenomenon and the consistency between the two relevant situations (with the first and second real-life phenomena). The interrelationship among these three correlations is illustrated in Fig. 7.8.

Another method is to confirm the feasibility of creating a more creative *design idea* by extrapolating an actual *design idea* on the basis of the virtual concept synthesis process. In this method, some of the features of a given *design idea* (hereafter, referred to as *base design idea*) are replaced by an equal number of *design idea features* from another given *design idea.* Specifically, the features with low polysemy of a *base design idea* are replaced by those with high polysemy of another *design idea.* Figure 7.9 illustrates this process of feature replacement. Each

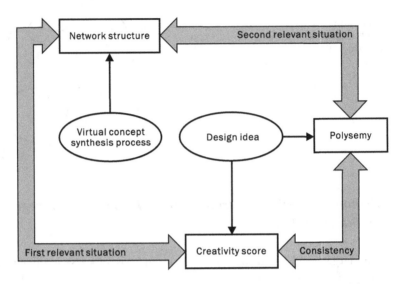

Fig. 7.8 Interrelationship among correlations between network structure, creativity score, and polysemy

circle represents a word which expresses a *design idea feature*, and each number represents its polysemy (number of meanings).

On the basis of the methods proposed above and in Sects. 7.2.2 and 7.2.3, we developed the following method ('polysemy-based replacement method') in order to create a set of *design idea features* which are expected to be more creative than the *base design idea* (Fig. 7.10).

- Step 1: Select some *design ideas* with high creativity—a *base design idea* and *design ideas* which will be combined with it—from the existing *design ideas*.
- Step 2: Using the developed method described above in this section, combine the *design idea features* of the *base design idea* and another *design idea* selected in Step 1 (creation).
- Step 3: Using the method developed in Sect. 7.2.2, a virtual concept synthesis process is constructed for each set of the *design idea features* created in Step 2. The creativity scores for each set are then estimated by referring to the criterion of the network structure of the newly constructed virtual concept synthesis process (estimation).

We refer to these two methods of verifying the virtual concept synthesis process (a constructive mechanism) as *synthetic verification*.

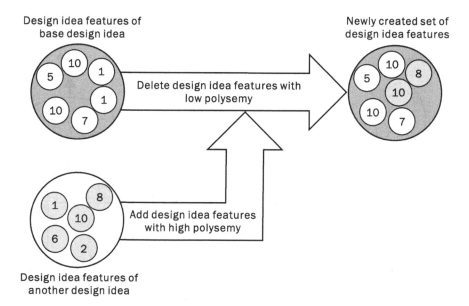

Fig. 7.9 Replacement of design idea features

7.3 Execution of Constructive Simulation for the Concept Synthesis

The method for *constructive simulation* was applied to data obtained from the design experiment introduced in Chap. 5.

7.3.1 Design Ideas Used for the Simulation

In the design experiment, the subjects were 22 undergraduate and graduate students majoring in industrial design. In the design task, they were required to design a new idea from two given *base concepts*. The *base concepts* used in this design task were 'ship-guitar' and 'desk-elevator'. The subjects came up with 20 *design ideas* for 'ship-guitar' and 19 for 'desk-elevator', which were adequate for the simulation. The creativity of these *design ideas* was evaluated from the viewpoint of practicality (whether it is achievable and feasible) and originality (whether it is novel), referring to the creativity evaluation method given in Finke et al. [13]. There were 11 raters who evaluated all the *design ideas* on a 5-point scale ranging from 1 (low) to 5 (high). Thereafter, the averages of the evaluated scores were calculated for each *design idea*.

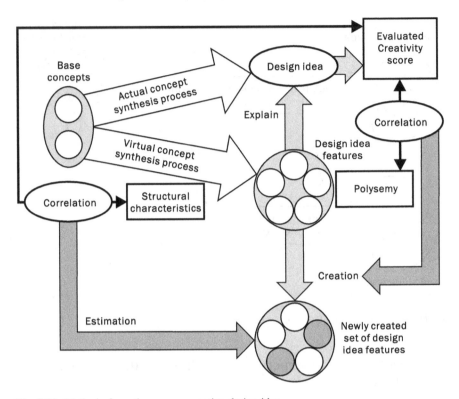

Fig. 7.10 Method of creating a more creative design idea

7.3.2 Construction of the Virtual Concept Synthesis Process

We constructed the virtual concept synthesis process for all the 39 *design ideas*. In WordNet, links are presented only between words which are structured in the same POS (part of speech). Thus, all verbs and adjectives are replaced by their corresponding noun forms. After translating the words into English and performing this preprocess, we constructed virtual concept synthesis processes according to the procedure explained in Sect. 7.2.2. Figure 7.11 presents two examples of virtual concept synthesis processes which were drawn using Pajek,[1] along with their nodes (words), the corresponding creativity scores (practicality and originality), and values of the network criteria.

---

[1] Program for large network analysis (version 1.23). http://vlado.fmf.uni-lj.si/pub/networks/pajek/. Accessed 24 December 2008.

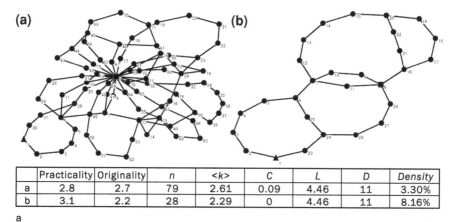

	Practicality	Originality	n	<k>	C	L	D	Density
a	2.8	2.7	79	2.61	0.09	4.46	11	3.30%
b	3.1	2.2	28	2.29	0	4.46	11	8.16%

a

Number of nodes	Label of nodes	Number of nodes	Label of nodes	Number of nodes	Label of nodes
1	base concepts	28	electronic network	55	thing
2	watercraft	29	system	56	channel
3	craft	30	stringed instrument	57	linguistic unit
4	vehicle	31	instrument	58	constituent
5	transport	32	falcon	59	auditory communication
6	instrumentation	33	hawk	60	communication
7	artifact	34	raptorial bird	61	natural event
8	design idea features	35	bird	62	mechanical phenomenon
9	textile	36	craniates	63	physical phenomenon
10	game equipment	37	chordate	64	natural phenomenon
11	equipment	38	fauna	65	phenomenon
12	goal	39	being	66	physical process
13	unit	40	animate thing	67	sense datum
14	physical object	41	toy	68	perception
15	physical entity	42	pursuit	69	basic cognitive process
16	entity	43	recreation	70	cognitive operation
17	income	44	activity	71	noesis
18	financial gain	45	human activity	72	sound property
19	gain	46	event	73	property
20	amount of money	47	psychological feature	74	attribute
21	assets	48	lepton	75	exercise
22	possession	49	fundamental particle	76	spouse equivalent
23	relation	50	fermion	77	soul
24	abstract entity	51	subatomic particle	78	causal agency
25	trap	52	body	79	follower
26	device	53	natural object		
27	computer network	54	water		

b

Number of nodes	Label of nodes	Number of nodes	Label of nodes	Number of nodes	Label of nodes
1	base concepts	11	article	21	quality
2	watercraft	12	unit	22	attribute
3	craft	13	physical object	23	stringed instrument
4	vehicle	14	physical entity	24	instrument
5	transport	15	entity	25	device
6	instrumentation	16	originality	26	clock
7	artifact	17	power	27	horologe
8	design idea features	18	noesis	28	measuring device
9	adornment	19	psychological feature		
10	ornamentation	20	abstract entity		

Fig. 7.11 Examples of the virtual concept synthesis process ('ship-guitar') with high originality (**a**) and low originality (**b**). ▲: base concept node; ■: design idea feature node

Table 7.2 Coefficients of correlation between the structural criteria and the evaluated creativity scores. Data include all 39 design ideas (20 for 'ship-guitar' and 19 for 'desk-elevator')

		n	$<k>$	C	L	D	*Density*
Average evaluated practicality score	Pearson correlation coefficient	0.058	0.133	0.005	−0.264	−0.222	−0.127
	Significance level	0.725	0.420	0.974	0.104	0.174	0.440
	Number of design ideas	39	39	39	39	39	39
Average evaluated originality score	Pearson correlation coefficient	0.290[+]	0.352[*]	0.103	−0.320[*]	−0.201	−0.398[*]
	Significance level	0.073	0.028	0.533	0.047	0.220	0.012
	Number of design ideas	39	39	39	39	39	39

+ p <0.10; * p <0.05; ** p <0.01

Fig. 7.12 Correlation between the evaluated originality score and $<k>$

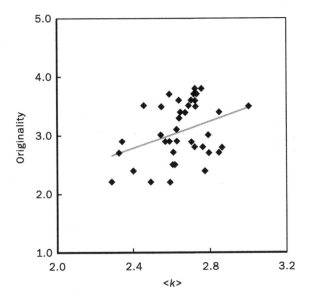

7.3.3 *Confirmation of the Relevance of the Virtual Concept Synthesis Process*

We examined the relationship between the structural criteria of the virtual concept synthesis process and the evaluated creativity scores of the *design ideas*.

As indicated in Table 7.2, $<k>$ had a significant positive correlation, with the evaluated originality score of $p < 0.05$, and *Density* had a significant negative correlation, with the evaluated originality score of $p < 0.05$. Scatter graphs depicting these respective correlations are presented in Figs. 7.12 and 7.13. In addition, n had a marginally significant positive correlation, with the evaluated originality score of $p < 0.1$ (Fig. 7.14).

Fig. 7.13 Correlation between the evaluated originality score and *Density*

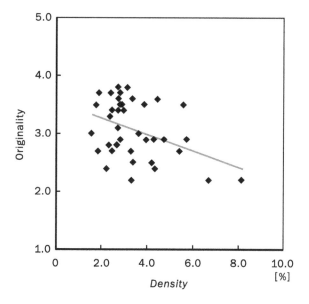

Fig. 7.14 Correlation between the evaluated originality score and *n*

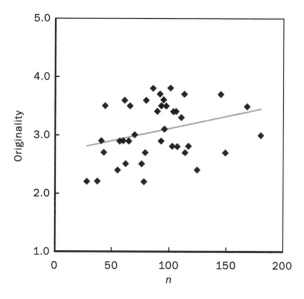

Further, L had a significant negative correlation, with an evaluated originality score of $p < 0.05$; the scatter graph is presented in Fig. 7.15.

The rank correlation test revealed a significant correlation of $p < 0.1$ for $<k>$ and L, which provided further evidence of a correlation between the two criteria and the evaluated score of originality. On the other hand, no significant

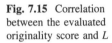

Fig. 7.15 Correlation between the evaluated originality score and L

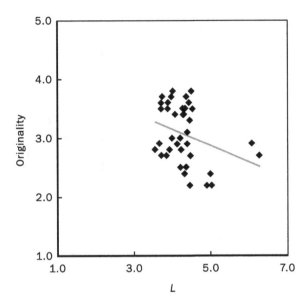

correlations were revealed for n ($p = 0.103$) or *Density* ($p = 0.121$). This absence of correlation could be ascribed to the weak correlation between n and the evaluated score of originality, which exerted an influence on *Density* (obtained by dividing $<k>$ by n). Thus, we conclude that $<k>$ and L have valid positive and negative significant correlations with the evaluated score of originality.

These results suggest that phenomenon—similar or analogous to the real-life phenomenon (actual *design ideas*)—emerged from the virtual concept synthesis process, thereby confirming the relevance of the constructed virtual concept synthesis process. However, no significant correlation was detected for the practicality score. This will be further discussed in the following section.

7.3.4 Synthetic Verification of the Virtual Concept Synthesis Process

7.3.4.1 Confirmation of the Relevance with Regard to the Second Real-Life Phenomenon

In preparation for *synthetic verification*, the polysemy (number of meanings) of all *design idea features* (words) was estimated using WordNet and the average of polysemy (average number of meanings) for each *design idea* was calculated.

First, we examined the correlations between the average of polysemy of the *design idea features* and the structural criteria of the virtual concept synthesis process. The results are presented in Table 7.3. We found that the average of

Table 7.3 Coefficients of correlation between the structural criteria and the average of polysemy of design ideas. Data include all 39 design ideas (20 for 'ship-guitar' and 19 for 'desk-elevator')

	n	$<k>$	C	L	D	Density
Pearson correlation coefficient	0.486**	0.371*	0.252	−0.229	−0.085	−0.534**
Significance level	0.002	0.020	0.122	0.161	0.606	0.000
Number of design ideas	39	39	39	39	39	39

$+ p < 0.10$; $* p < 0.05$; $** p < 0.01$

Fig. 7.16 Correlation between the average of polysemy of the corresponding design idea and n

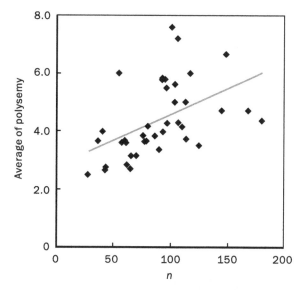

polysemy of the *design ideas* had positive correlations with n and $<k>$ ($p < 0.01$ and $p < 0.05$, respectively; Figs. 7.16 and 7.17, respectively), and a negative correlation with *Density* ($p < 0.01$; Fig. 7.18). These results confirm the relevance of the virtual concept synthesis process with regard to the second real-life phenomenon.

Next, the correlation between the average of polysemy of the *design ideas* and the creativity score of the *design ideas* was examined. The results are presented in Table 7.4. We found that the average of polysemy of the *design idea* had a significant positive correlation with the average evaluated originality score of the corresponding *design idea* ($p < 0.05$; Fig. 7.19), while no correlation was found with the average evaluated practicality score ($p = 0.836$).

This correlation confirms the consistency between the two relevant situations—one with the evaluated originality score (the first real-life phenomenon) and the other with the polysemy (the second real-life phenomenon)—even though the data of these two real-life phenomena were obtained in different ways. The former was obtained from the raters, while the latter was obtained from the *design idea*

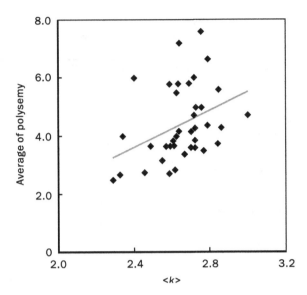

Fig. 7.17 Correlation between the average of polysemy of the corresponding design idea and <*k*>

features described by the subjects in order to explain their *design ideas*. Therefore, this consistency supports the validity of the virtual concept synthesis process.

It must be noted that no significance could be found in the scores for practicality. It is believed that the richness of the polysemy of a *design idea* is not strongly related to its practicality since this richness is not assumed to have a direct connection with an idea's achievability or feasibility. In Sect. 7.3.3, no significant correlation was found between the practicality score and the structural criteria of the virtual concept synthesis process. This absence of correlation also supports the consistency between the two relevant situations.

7.3.4.2 Confirmation of Feasibility of Creating More Creative Design Ideas

We conducted a case study to create more creative *design ideas* in which we used the *design ideas* for 'ship-guitar' as the existing *design ideas*. First, the regression equation for the evaluated originality score and the number of nodes *n* of the virtual concept synthesis process of the *design ideas* for 'ship-guitar' was obtained (Fig. 7.20); then, it was used for evaluation of the newly created set of *design idea features*.

Thereafter, we selected the *design ideas* which had the four highest scores for originality, using the regression equation shown in Fig. 7.20. We selected B6 as the *base design idea* and A2, B5, and B9 as the *design ideas* to be combined with it. Table 7.5 presents the values for each *design idea*: the evaluated originality score, originality score as calculated using the regression equation, number of nodes in the virtual concept synthesis process, and number of *design idea features*.

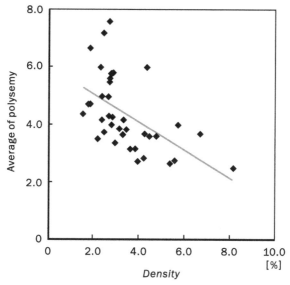

Fig. 7.18 Correlation between the average of polysemy of the corresponding design idea and *Density*

Table 7.4 Correlation coefficients between the average of polysemy and the evaluated creativity scores. Data include all the 39 design ideas (20 for 'ship-guitar' and 19 for 'desk-elevator')

Evaluated practicality score	Pearson correlation coefficient	0.034
	Significance level	0.836
	Number of design ideas	39
Evaluated originality score	Pearson correlation coefficient	0.345[*]
	Significance level	0.032
	Number of design ideas	39

$+ p < 0.10$; $* p < 0.05$; $** p < 0.01$

The selected *design idea features* and their polysemy (number of meanings) are presented in Table 7.6. We combined the *design idea features* using the polysemy-based replacement method described earlier. In order to verify the effectiveness of this method, we also combined *design idea features* using a random replacement method.

Figure 7.21 illustrates the virtual concept synthesis process for the *base design idea* (left) and the newly created set of *design idea features* (right). Figures 7.22 and 7.23 illustrate the originality scores for each newly created set of *design idea features* as obtained using the polysemy-based replacement method and random replacement method, respectively. The scores were estimated by following Step 3 in Sect. 7.2.4, using the regression equation shown in Fig. 7.20. The circles indicate the calculated originality scores of the *base design idea* (B6).

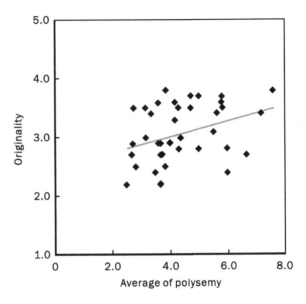

Fig. 7.19 Correlation between the evaluated originality score of design idea and the average of polysemy of the corresponding design idea

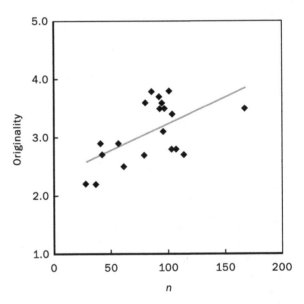

Fig. 7.20 Correlation between the evaluated originality score and *n* for 'ship-guitar'

Thereafter, we attempted to draw a picture on the basis of the newly created set of *design idea features* which had the highest originality scores. The new set of *design idea features* was obtained by replacing "pleasantness", "symbiosis", "humankind", and "egalitarianism" with "string", "shape", "performance", and "material"; respectively, from the *design idea features* of B5. Figure 7.24 illustrates the resulting sketch drawn by us. This *design idea* represents "water-borne

Table 7.5 Selected design ideas

	Evaluated originality score	Estimated originality score	Number of nodes	Number of design idea features
B6	3.5	3.9	168	18
A2	2.7	3.4	114	12
B5	2.8	3.3	107	10
B9	3.4	3.3	104	8

vehicle driven by a duck, dolphin, or fish-shaped robot which is manipulated by the sound generated when a string is plucked by a person in the vehicle".

The estimated originality scores of the newly created set of *design idea features* obtained using the polysemy-based replacement method to extrapolate one of the *base design ideas* are evident in Fig. 7.22. These scores were estimated by calculating the criteria of the network structure of the virtual concept synthesis process. These results suggest that the polysemy-based replacement method, which was used to create a new set of *design idea features*, is consistent with the method used to estimate the originality scores of *design ideas* by referring to the network structure criterion of the virtual concept synthesis process.

We consider that the consistency of the results of the virtual concept synthesis process and polysemy-based replacement method imply that there is consistency between the originality score, polysemy, and virtual concept synthesis process; thus, we indicate the validity of the conducted *constructive simulation*.

Furthermore, we consider that the new *design idea* visualized from the new set of *design idea features* (Fig. 7.24) is a creative idea of distinct originality. Indeed, the very fact that we could actually design a new idea using methods based on the virtual concept synthesis process can be considered not just a verification of the validity of these methods from another perspective, but an evidence of their implementability as well.

7.3.5 Inference of the Characteristics of the Actual Concept Synthesis Process or Patterns from the Virtual Concept Synthesis Process

The characteristics of the virtual concept synthesis process are as follows. The same semantic network was used for each designer. The process was represented as a continuous network composed of the paths between *base concepts* and *design idea features* which contain virtual concepts. These paths were searched at the abstract level; the parameters $<k>$ and L showed a significant correlation with the average score for originality, and explicitly expressed 'words' were used for the simulation. From these characteristics, the characteristics of the actual concept synthesis process can be inferred.

Table 7.6 Design idea features and the polysemy of each design idea (number of polysemy is shown in parentheses)

B6	A2	B5	B9
Field (17)	Picture (10)	String (10)	Tone (10)
Land (11)	Wood (8)	Shape (8)	Sound (8)
First (6)	Scrap (4)	Performance (5)	Shape (8)
Water (6)	Room (4)	Music (5)	Function (7)
Toy (5)	Plant (4)	Material (5)	Feature (6)
Practice (5)	Fashion (4)	Thickness (4)	Beauty (3)
Peace (5)	Deck (4)	Electricity (3)	Originality (2)
Pleasure (5)	Planter (3)	Ship (1)	Ship (1)
Fish (4)	Recycling (1)	Parts (1)	
Novelty (4)	Floral arrangement (1)	Recycling (1)	
Brightness (3)	Décor (1)		
Imagination (3)	Corkboard (1)		
Experience (3)			
Enjoyment (3)			
Pleasantness (2)			
Symbiosis (1)			
Humankind (1)			
Egalitarianism (1)			

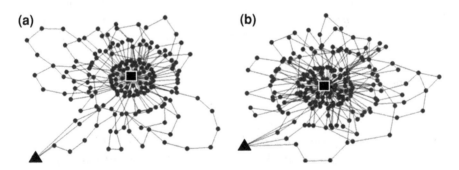

Fig. 7.21 Examples of the virtual concept synthesis process for the base design idea (**a**) and a newly created set of design idea features (**b**). ▲: base concept node; ■: design idea feature node

First, we used the same semantic network to construct a virtual concept synthesis process for each designer. This suggests that the creativity of the *design ideas* (*design idea features*) produced by individual designers does not originate from the differences in the structures of the concepts in their minds; rather, it originates from the manner in which their thinking proceeds. Accordingly, this suggests the

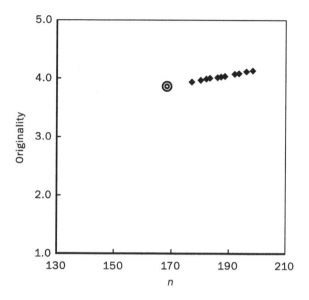

Fig. 7.22 Estimated originality scores for each set of design idea features created by polysemy-based replacement. ⊙: base design idea; ◆: newly created set of design idea features

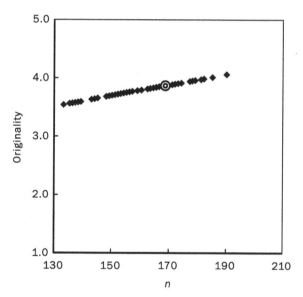

Fig. 7.23 Estimated originality scores for each set of design idea features created by random replacement. ⊙: base design idea; ◆: newly created set of design idea features

possibility that there are common characteristics or patterns in the process of *concept synthesis* which effectively lead to a highly creative *design idea*.

Second, the virtual concept synthesis process involves the notion of 'continuity', which is composed of the virtual concepts. This leads to the following discussions. In cognitive psychology, since the 1970s, there has been an intense ongoing debate on the subject of implicit cognition [24]. It was confirmed that no obvious

Fig. 7.24 Sketch of a new design idea based on the set of design idea features with the highest estimated originality scores

borderline separates implicit processes from explicit knowledge, and a sort of border zone, referred to as 'inexplicit' cognition and which is located somewhere between explicit and implicit cognition, has been recognized. For example, language is understood to be derived from both explicit and inexplicit cognition [26], and as a result, inexplicit language has been employed as a means of investigating the nature of language [16]. It is apparent that human beings are unable to express their every thought in an explicit manner. However, even though it may be difficult to precisely express all thoughts using words, humans might be capable of recognizing certain elements of their thoughts which are expressible. In the virtual concept synthesis process, certain virtual concepts are believed to correspond to certain inexplicit concepts in an actual concept synthesis process, and these concepts are inexplicitly recognized; under certain circumstances, these concepts may also be explicitly recognized by the designer. Thus, we can infer that 'inexplicit' cognition in the process of *concept synthesis* plays an important role in the production of a creative *design idea*. On the other hand, our method of representing the virtual concept synthesis process as a network can be integrated with the idea that the association process between one concept and another constitutes the basis of the thinking process in design, which was mentioned in Chap. 6. Consequently, we can re-assert that *concept synthesis* involves a network of associative thought. While the phenomenon of leaps of creative thinking in design has been widely noted (e.g. [6]), an alternative theory advocates the idea that reasoning proceeds in increments [3, 32]. In this regard, the results of our simulation imply that the creative thinking process is originally continuous, but it appears to be noncontinuous because it is partially hidden within the 'inexplicit' mind.

Third, the paths were searched at the abstract level. As mentioned in Sect. 7.2.2, the route between the *base concept* and *design idea features* was determined by searching for all nodes on the upper layer of *base concept* meanings and *design idea feature* meanings, respectively. This process implies that every route is searched at a more abstract level than the *base concepts* or *design idea features*, and it is inferred that this abstraction plays an important role in the actual concept synthesis process. This inference is consistent with the existing literature, thus highlighting the importance of abstraction in the design process [2, 20, 22, 31].

Fourth, the statistical results of our analysis of the virtual concept synthesis process suggest the existence of thinking patterns in the creative *concept synthesis* process. In particular, the fact that the parameter $<k>$ and L showed a significant correlation with the average score for originality implies that the thinking patterns in which both explicit and inexplicit concepts are 'intricately intertwined' in a complex manner may lead to a creative *design idea*. This inference is consistent with the existing literature which investigates the characteristics of the design thinking process. For example, $<k>$ corresponds to the 'critical moves' in 'linkography' [15] and the complexity of a network is considered to be consistent with the notion of 'entropy' in Kan and Gero [17].

Finally, we must also address the fact that we used words in our quest to deal with difficult-to-observe phenomena. This presents a paradox, since words are one of the most readily observed forms of information. In this simulation, it is assumed that inexplicit concepts are dealt with as described above. It is further assumed that the structure of the virtual concept synthesis process and the polysemy of the words explaining the *design ideas*, which cannot be observed directly, are extracted. Thus, it is believed that there are definite characteristics which lay hidden in observed words.

7.4 Towards the Dynamic Simulation

The procedure formulated in this simulation is classified as an 'ex post facto' simulation. In order to elucidate a process of *concept generation*, it is necessary to simulate a process which truly generates creative *design ideas*. For example, in the proposed method, if we switch the *base concepts* and *design idea features* (i.e. use *design idea features* as the starting point and *base concepts* as the ending points), exactly the same process will be produced. Such a result is possible because this method does not consider the factor of time polarity. In other words, the simulation is, as it were, a 'static' event. In order to develop this process further, we would need to conduct a more 'dynamic' simulation which takes into account the factor of time polarity in thinking.

One way to take this next step would be to use a semantic network which includes time polarity. An example of this is the Associative Concept Dictionary

[23], in which the causal relationships between stimuli and responses are descri-bed. However, this particular dictionary is not yet applicable to the present chal-lenge owing to the limited number of lexical entries in the currently available version.

Another approach would be to conduct a generation process which proceeds sequentially from a starting point and is governed by certain rules of generation. However, for now, the rules which could govern the way humans produce *design idea features* remain a subject which needs clarification. We can speculate that such clarification might even reveal the 'dynamic' features of the design thinking process which generates creative *design ideas*. Such a discovery could also have important applications in the field of engineering, where the generation of creative *design ideas* is often a significant activity.

References

1. Amabile TA (1996) Creativity in context: update to the social psychology of creativity. Westview Press, New York
2. Andreasen MM (1994) Modelling—the language of the designer. J Eng Des 5:103–115
3. Brown DC (2010) The curse of creativity. In: Gero JS (ed) Design computing and cognition '10. Springer, London
4. Cancho RFI, Solé RV (2001) The small world of human language. Proc R Soc Lond 268:2261–2265. doi:10.1098/rspb.2001.1800
5. Chiu I, Shu LH (2007) Using language as related stimuli for concept generation. AI EDAM 21:103–121. doi:10.1017/S0890060407070175
6. Cross N (2006) Designerly ways of knowing. Birkhäuser, Basel
7. Coyne RD, Newton S, Sudweeks F (1993) A connectionist view of creative design reasoning. In: Gero JS, Maher ML (eds) Modeling creativity and knowledge-based creative design. Hillsdale, New Jersey
8. Csikszentmihalyi M (1990) Flow: the psychology of optimal experience. Harper & Row, New York
9. Dong A (2006) Concept formation as knowledge accumulation: a computational linguistics study. AI EDAM 20:35–53. doi:10.1017/S0890060406060033
10. Ericsson KA, Simon HA (1984) Protocol analysis: verbal reports as data. The MIT Press, Cambridge
11. Fauconnier G, Turner M (2003) Polysemy and conceptual blending. In: Nerlich B, Todd Z, Herman V, Clarke DD (eds) Polysemy: flexible patterns of meaning in mind and language. Mouton de Gruyter, Berlin
12. Fellbaum C (1998) WordNet: an electronic lexical database. The MIT Press, Cambridge
13. Finke RA, Ward TB, Smith SM (1992) Creative cognition: theory, research, and applications. The MIT Press, Cambridge
14. Georgiev GV, Taura T, Chakrabarti A, Nagai Y (2008) Method of design through structuring of meanings. In: Proceedings of ASME 2008 international design engineering technical conference and computers and information in engineering conference. Brooklyn, New York, 3–6 August (CD-ROM)
15. Goldschmidt G (1990) Linkography: assessing design productivity. In: Trappl R (ed) Cybernetics and systems 90. World Scientific, Singapore, pp 291–298
16. Higginbotham J (2002) On linguistics in philosophy, and philosophy in linguistics. Ling Philos 25:573–584. doi:10.1023/A:1020891111450

17. Kan JWT, Gero JS (2008) Acquiring information from linkography in protocol studies of designing. Des Stud 29:315–337. doi:10.1016/j.destud.2008.03.001
18. Kokotovich V (2008) Problem analysis and thinking tools: an empirical study of non-hierarchical mind mapping. Des Stud 29:49–69. doi:10.1016/j.destud.2007.09.001
19. Loewenstein G (1994) The psychology of curiosity: a review and reinterpretation. Psychol Bull 116:75–98. doi:10.1037/0033-2909.116.1.75
20. Nagai Y, Noguchi H (2004) An experimental study on the design thinking process started from difficult keywords: modelling the thinking process of creative design. J Eng Des 14:429–437. doi:10.1080/09544820310001606911
21. Nagai Y, Taura T (2010) Discussion on direction of design creativity research (part 2)—research issues and methodologies: from the viewpoint of deep feelings and desirable figure. In: Taura T, Nagai Y (eds) Design creativity 2010. Springer, London
22. Nagel RL, Hutcheson R, McAdams DA, Stone R (2011 online 2009) Process and event modelling for conceptual design. J Eng Des 22:145–164. doi:10.1080/09544820903099575
23. Okamoto J, Ishizaki S (2001) Associative concept dictionary construction and its comparison with electronic concept dictionaries. In: Proceedings of the Pacific Association for computational linguistics conference 2001. Kitakyushu, Japan, 11–14 Sept, pp 214–220
24. Reingold EM, Ray CA (2002) Implicit cognition. In: Nadel L (ed) Encyclopedia of cognitive science. Nature Publishing Group, London
25. Schilling MA (2005) A 'small-world' network model of cognitive insight. Creativ Res J 17:131–154. doi:10(1080/10400419),2005,9651475
26. Searle J (1975) Indirect speech acts. In: Cole P, Morgan JL (eds) Syntax and semantics, vol 3. Speech Acts. Academic Press, New York
27. Steyvers M, Tenenbaum JB (2005) The large-scale structure of semantic networks: statistical analyses and a model of semantic growth. Cognit Sci 29:41–78. doi:10.1207/s15516709cog2901_3
28. Taura T, Yamamoto E, Fasiha MYN, Goka M, Nagai Y, Nakashima H (2011) Trial of a constructive research method for the creative thinking process in design—constructive simulation of the concept generation process. Cognit Stud 18:329–341 (in Japanese)
29. van der Lught R (2002) Functions of sketching in design idea generation meetings. In: Proceedings of the 4th conference on creativity & cognition. Loughborough, UK, 13–16 Oct, pp 72–79. doi: 10.1145/581710.581723
30. Varela FJ, Thompson E, Rosch E (1997) The embodied mind: cognitive science and human experience. The MIT Press, Cambridge
31. Ward TB, Patterson MJ, Sifonis CM (2004) The role of specificity and abstraction in creative idea generation. Creativ Res J 16:1–9. doi:10.1207/s15326934crj1601_1
32. Weisberg RW (1986) Creativity: genius and other myths. WH Freeman and Company, New York

Chapter 8
Constraints in Concept Synthesis: Distance Between and Association of Base Concepts

Abstract In this chapter, the constraints for the *base concepts* in generating a creative *design idea* in *concept synthesis*, particularly in *concept blending*, are discussed. We conduct two experiments to determine the conditions, by focusing on the distance between the *base concepts* and the *associative concepts* of the *base concepts*. The findings of the former experiment reveal that the highest evaluated score of originality was obtained in the case of blending the *base concepts* with high distance, while those of the latter experiment reveal that the *base concepts* with a high number of *associative concepts* lead to high originality in designed outcomes.

8.1 Distance Between the Base Concepts

In this section, we introduce an experiment to determine the conditions for successful *concept synthesis* by focusing on the distance between the *base concepts*. The reason we pay attention to the distance between the *base concepts* is because the 'similarity-recognition process' and 'dissimilarity-recognition process' are important elements in the *concept generation*, as mentioned in Chap. 4, and 'similarity' and 'dissimilarity' comprise the notion of 'distance' between concepts. In particular, the 'dissimilarity-recognition process' is considered to be related to *high-order concept generation*, which is expected to lead to an *innovative design idea* beyond the existing category. In this experiment, the effect of the distance between the *base concepts* on the creativity of the designed outcomes is examined in *high-order concept generation*, particularly in *concept blending*.

T. Taura and Y. Nagai, *Concept Generation for Design Creativity*,
DOI: 10.1007/978-1-4471-4081-8_8, © Springer-Verlag London 2013

8.1.1 Method of the Experiment

8.1.1.1 Preliminary Experiment to Investigate the Subjects' Individual Recognition of the Distance Between the Entity Concepts

The distance between concepts is recognized differently by each person; hence, it was measured for every subject at the beginning of the experiment. Concretely, the subjects were required to arrange nine *entity concepts*, which were selected by the experimenters, from among some artificial and natural objects—"plastic bottle", "charcoal", "fish", "bird", "star", "thermometer", "scissors", "flower", and "wind"—according to the distance from the viewpoint of function and form from "glass", which was the standard *entity concept* in this experiment. The nine *entity concepts* were divided into three groups with respect to the distance of function and form: low-distance, intermediate-distance, and high-distance groups. Considering this result, the *base concepts* used in the design tasks were selected from the three groups.

8.1.1.2 Design Tasks

Before performing the design tasks, a preparatory experiment (10 min) was carried out in order to help the subjects get accustomed to the *concept blending*. The subjects were given an explanation about the blending process of the *base concepts* by using examples. They were also instructed that they had to better abstract the two *base concepts* by extracting some features from both as long as the final designed outcome inherited a certain feature from each. The task was to design a new idea from "car" and "dolphin". After the preparatory experiment, three design tasks (10 min each) were assigned to each subject in turn.

These three tasks were as follows.

- First design task (10 min): Design a new idea by blending "glass" and another *base concept* selected from the low-distance group.
- Second design task (10 min): Design a new idea by blending "glass" and another *base concept* selected from the intermediate-distance group.
- Third design task (10 min): Design a new idea by blending "glass" and another *base concept* selected from the high-distance group.

8.1.1.3 Subjects

The participants in the experiments comprised two subjects who were undergraduate students majoring in industrial design.

Table 8.1 Distance from "glass"

Subject	View point	Low	Intermediate	High
S	Form	**Plastic bottle**	Star	Fish
		Flower	Wind	**Bird**
		Thermometer	**Charcoal**	Scissors
	Function	**Plastic bottle**	Star	Fish
		Flower	**Charcoal**	Scissors
		Wind	Thermometer	**Bird**
F	Form	Flower	**Fish**	Charcoal
		Plastic bottle	Thermometer	Scissors
		Star	Wind	**Bird**
	Function	**Plastic bottle**	**Fish**	**Bird**
		Flower	Scissors	Wind
		Thermometer	Star	Charcoal

8.1.1.4 Creativity Evaluation

The creativity of the *design ideas* was evaluated from the viewpoint of practicality (whether the idea is achievable and feasible) and originality (whether the idea is novel), referring to the creativity evaluation method given in Finke et al. [2]. In all, 10 raters evaluated all the *design ideas* on the basis of a five-point scale (1: low and 5: high). The rating scores were averaged for each *design idea*.

8.1.2 Results

Table 8.1 presents the result of the preliminary experiment investigating the subjects' individual recognition of the distance between the *entity concepts*. Considering this result, "plastic bottle" for low distance with "glass" and "bird" for high distance with "glass", which are the same between the two subjects, were selected. The *base concept* of intermediate distance was different between the two subjects; hence, "charcoal" and "fish" were selected for each subject. Therefore, the *design idea* which was obtained by blending "glass" and "plastic bottle" was the case of low distance between the two *base concepts*, and that of "glass" and "bird" was the case of high distance between the *base concepts*. Figure 8.1 illustrates the *design ideas* generated during the design tasks by the two subjects.

Table 8.2 shows the evaluated creativity score of each *design idea*. Kendall's coefficient of concordance showed a significant concordance in both originality and practicality (originality: $W = 0.48$, χ^2 (5) = 23.76, $p < 0.01$; practicality: $W = 0.25$, χ^2 (5) = 12.56, $p < 0.05$). The highest score of practicality was obtained by F2 (3.5), which was the *design idea* obtained by blending "glass" and "fish". The highest evaluated score of originality was obtained by F3 (4.3), which was the *design idea* generated by blending "glass" and "bird". The lowest evaluated score of practicality was also obtained by F3 (1.7). The lowest evaluated

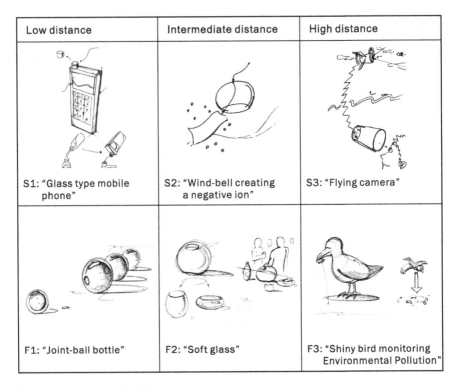

Low distance	Intermediate distance	High distance
S1: "Glass type mobile phone"	S2: "Wind-bell creating a negative ion"	S3: "Flying camera"
F1: "Joint-ball bottle"	F2: "Soft glass"	F3: "Shiny bird monitoring Environmental Pollution"

Fig. 8.1 Generated design ideas

Table 8.2 Evaluated creativity score

	S1	S2	S3	F1	F2	F3
Practicality	3.3	3.2	3.2	2.7	3.5	1.7
Originality	3.5	1.8	4.0	3.5	3.5	4.3
Distance	Low	Middle	High	Low	Middle	High

score of originality was obtained by S2 (1.8), which was generated by blending "glass" and "charcoal". The highest evaluated score of originality was obtained in the third design task for both subjects.

The results revealed that the highest evaluated score of originality was obtained in the case of blending the *base concepts* with high distance.

8.2 Association of the Base Concepts

In this section, we introduce an experiment to determine the conditions for successful *concept blending* by focusing on the number of *associative concepts*. An *associative concept* of the concept A is one which is associated with concept

Table 8.3 Number of associative concepts of the 20 selected entity concept

Artificial objects	Number of associations	Natural objects	Number of associations
Mirror	114	Flower	151
Glasses	110	Dog	127
Bag	97	Fish	115
Letter	93	Bird	112
Chair	91	Milk	91
Scissors	81	Water	85
Pool	75	Oil	78
Guitar	74	Egg	71
Blanket	69	Star	64
Thermometer	68	Ice	52

A. The reason we pay attention to the number of *associative concepts* is that the association process is expected to activate *concept synthesis*, which was pointed out in Chaps. 6 and 7, even more if the *base concepts* are rich in *associative concepts*.

8.2.1 Method of the Experiment

8.2.1.1 Design Tasks

First, 20 *entity concepts* were selected by the experimenters from the artificial and natural objects groups. The selected *entity concepts* were "mirror", "glasses", "bag", "letter", "chair", "scissors", "pool", "guitar", "blanket", "thermometer", "flower", "dog", "fish", "bird", "milk", "water", "oil", "egg", "star", and "ice".

Next, the number of *associative concepts* from each *entity concept* was counted by using the Associative Concept Dictionary [3]. The Associative Concepts Dictionary is an electronic dictionary which was developed from lists of large numbers of words evoked from the selected words from a fundamental vocabulary [4]. Table 8.3 shows the number of *associative concepts* of the 20 selected *entity concepts*. In addition, we measured the distances between the selected *entity concepts* by using the EDR Concept Dictionary [1], introduced in Chap. 6, in order to select *base concepts* with equal distances so that the effects of distance on creativity can be excluded. To determine the distances between concepts, we used the same method to count the least number of steps between two concepts in the EDR Concept Dictionary.

Pairs of *entity concepts* with equal distances were selected as the *base concepts* for the design tasks. For Task A, "egg" and "blanket", which had a low number of *associative concepts*, were selected. For Task B, "flower" and "mirror", which had a high number of *associative concepts*, were selected. "Vehicle" was selected as the target category of the design because it was equidistant from "egg", "blanket", "flower", and "mirror".

The design tasks assigned to the subjects in this experiment were as follows.

- Task A: Design a new vehicle by blending "egg" and "blanket" (low number of *associative concepts*).
- Task B: Design a new vehicle by blending "flower" and "mirror" (high number of *associative concepts*).

8.2.1.2 Subjects

In all, five subjects participated in the experiments: two were graduate students not majoring in industrial design, while two were graduate students majoring in industrial design; one was a professional industrial designer.

8.2.1.3 Procedure of the Experiment

Before performing the design tasks, a preparatory experiment (5 min) was carried out in order to help the subjects get accustomed to the *concept blending*. The subjects were given an explanation about the blending process of the *base concepts* by using examples. They were also instructed that they had to better abstract the two *base concepts* through extracting some features from them as long as the final designed outcome inherited a certain feature from each. The task involved designing a new idea by blending two *base concepts* ("glass" and "bird"). These two concepts were selected from a previous study, which is mentioned in Sect. 8.1.

In the experiment, the subjects were required to perform Tasks A and B (to control the sequence effort of the task order, subjects 1, 3, and 5 began with Task A and subjects 2 and 4 began with Task B) in 10 min, respectively.

8.2.1.4 Creativity Evaluation

The creativity of the designed outcomes was evaluated from the viewpoints of practicality and originality, referring to the creativity evaluation method given in Finke et al. [2]. In all, 10 raters evaluated all the *design ideas* on the basis of a four-point scale (1: low and 4: high). The rating scores were averaged for each designed outcome.

8.2.2 Results

Because some subjects were not experienced designers, creativity was evaluated on the basis of the content of the *design ideas* in order to avoid the influence of the drawing skill. The experimenter prepared the summaries (*design concept*) of the *design ideas*. The *design concepts* of each task are shown below.

- Subject 1

  - Task A: "Equipment of attraction which resembles a Gashapon (capsule toy)[1]"
    This is a playground equipment which looks like a toy capsule. There are four rails to hold and riders can slide down safely on the rails. The capsule's interior is covered with cushions.
  - Task B: "Stealth car which uses biomass energy"
    High technology is employed to protect the object from being detected or seen. This car runs on biomass energy.

- Subject 2

  - Task A: "Tricycle with a tatami"
    A blanket is spread across a 'tatami' (a Japanese mat) and people can ride a blanket-covered tricycle. There are three wheels under the tatami, and the tricycle is powered by an electric motor. Handling is easy and riders can lie down on the bike comfortably.
  - Task B: "Car which shrivels up if driven roughly"
    Good drivers can nurture this car to make it more beautiful. 'Good driving' refers not to the driving technique but to a person's concern for others. However, if driven roughly, the car will shrivel up.

- Subject 3

  - Task A: "Car which looks like an egg"
    The bonnet of the car is transparent, and since the car is shaped like an egg, it can be opened halfway, like an egg. A driver can enter the car through this opening. The interior of the car is covered with fur. It is pure white and can carry a maximum of two people.
  - Task B: "Tanker (a type of ship) with a huge planter"
    This tanker has a huge planter on top where flowers are cultivated. There is a mirror above the planter so that sunlight is always reflected onto the planter.

- Subject 4

  - Task A: "Egg slider attraction"
    Riders can go inside the egg and slide down a half pipe shaped like half an egg. The inside of the egg is covered in soft material so that the rider can lie down. The egg has a double-layered structure which works like a gyroscope to enable people to lie down comfortably while sliding.
  - Task B: "Kaleidoscopic train or bus"
    There are two sofas in the coach of a train or bus. They face each other and have mirrors behind them. These mirrors also face each other. There are

---

[1] http://www.gashapon.jp/ Accessed 12 December 2011.

Table 8.4 Evaluated score of creativity

	Practicality		Originality	
	Task A	Task B	Task A	Task B
Subject 1	2.4	2.3	2.0	2.0
Subject 2	1.4	2.4	2.0	3.1
Subject 3	2.1	2.3	1.7	2.5
Subject 4	2.4	2.3	1.7	2.4
Subject 5	2.1	2.4	2.2	2.3
Average (SD)	2.08 (0.41)	2.34 (0.05)	1.92 (0.22)	2.46 (0.40)

windows beside the sofas, and when the train or bus passes through a field of flowers, it creates a kaleidoscopic image which constantly changes with the reflection.

- Subject 5

  - Task A: "Flying Love Board"
    This is a disk-shaped board, about 10 cm in diameter, which can fly. The centre of the disk glitters and presents words or photographs. Young lovers enjoy riding on the disk and recording their conversation. They can play back their conversation when they push the glittering part of the disk.
  - Task B: "Cylindrical relaxation elevator"
    This elevator does not transfer people from one floor to another, but instead is used to view oneself as a user in a slow-moving space. Water falls on the surface of the cylinder every 3–4 min. People are relaxing in this elevator. A personal gondola goes up and down slowly in the cylinder and the user can sleep, think, meditate, or look outside (about 100 m height and 1.5 m diameter).

8.2.2.1 Creativity Evaluation of Design Concepts

The *design concepts* were evaluated by 9 raters (2 of whom were professional designers, while the rest were design professors). Table 8.4 summarizes the results of the evaluation. Kendall's coefficient of concordance showed a significant concordance in originality, whereas no significant concordance in practicality (originality: $W = 0.252$, χ^2 (9) = 20.49, $p < 0.05$; practicality: $W = 0.163$, χ^2 (9) = 13.27, n.s.).

The result shows that the evaluated score of originality for subjects 2, 3, and 4 are higher for Task B than Task A, while only subject 1 obtained the same score in both Tasks A and B. The two-sided t-test shows that there is a significant difference between Tasks A and B in terms of the evaluated score of originality (two-sided test: $t(4) = 2.55$, $p < 0.05$). However, there was no significant difference between Tasks A and B for the practicality.

These results suggest that the *base concepts* with a high number of *associative concepts* lead to higher originality in designed outcomes in *concept blending*.

References

1. EDR Concept Dictionary (2005) EDR Electronic dictionary. National Institute of Information and Communications Technology, CPD-V030 (CD-ROM)
2. Finke RA, Ward TB, Smith SM (1992) Creative cognition: theory, research, and applications. The MIT Press, Cambridge
3. Ishizaki S (2007) Associative concept dictionary (Ver. 2). Keio University, Fujisawa (CD-ROM, in Japanese)
4. Okamoto J, Ishizaki S (2001) Associative concept dictionary construction and its comparison with electronic concept dictionaries. In: Proceedings of the Pacific association for computational linguistics conference 2001. Kitakyushu, Japan, 11–14 Sept, pp 214–220

Chapter 9
Synthesis of Abstract Shape: Practice of Concept Generation (1)

Abstract In this chapter, a method to synthesize the abstract shapes is developed. This method is characterized by operating the abstract shapes on the space in which the abstract shapes are represented, by using the evaluation function from which the shape can be deduced using the optimization method. This method is composed of three steps. First, the evaluation function is acquired by using Genetic Algorithms (GA) and Genetic Programming (GP). Second, the acquired evaluation functions are combined. Third, a new abstract shape is created by using GA. In the experiment, the images of "chair" and "pyramid" are synthesized.

9.1 Blending of Abstract Shapes

Producing a creative abstract shape is an essential process to generate a creative visage of objects at the very early stage in the industrial design. In particular, the features of an abstract shape (hereafter, referred to as 'abstract form features') play an important role. However, such an abstract form feature is difficult to operate on a computer. For example, it is difficult to synthesize the abstract form features of "round" and "triangle" on the computer when retaining their abstract form features, but it is easy for humans to do so (Fig. 9.1). In order to support the generation of a creative abstract shape efficiently, it is necessary to develop a method to operate the abstract form features on the computer. Generating a new abstract shape by blending some abstract shapes (hereafter, referred to as *base shapes*) can be one such method.

9.2 Method of Blending the Abstract Form Features

To develop a method to operate the abstract form features, we convert the space in which the abstract shapes are described into another space in which the abstract form features can be operated (blended), because the abstract form features cannot be

T. Taura and Y. Nagai, *Concept Generation for Design Creativity*,
DOI: 10.1007/978-1-4471-4081-8_9, © Springer-Verlag London 2013

Fig. 9.1 Examples of
synthesis of "round" and
"triangle"

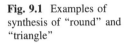

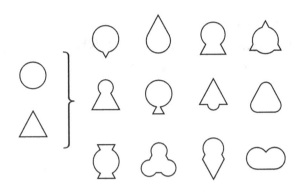

combined as is. Even if we attempt to combine the two abstract form features in the geometric space, the expected abstract shape may not be created; instead, another type of abstract shape will be created. It is difficult for the computer to understand and operate the abstract form features and synthesize them, and thus conform with the designer's image. In our approach, we pay attention to the evaluation function. Generally, an evaluation function implies a mathematical function for evaluating how an object matches the given purpose. Here it should be noted that the evaluation function can deduce the optimum shape. For example, a circle is usually described as the set of points whose distance to the centre is the same in the geometric space; instead, a circle can also be represented as the line whose circumference is shortest under the condition that the area within the line is given.

By extending this idea, we expect to be able to obtain any shape as the optimum shape from a certain evaluation function. On the basis of the above consideration, in this method, the evaluation function is used to represent the abstract shape. If the abstract shapes are represented by using the evaluation functions, the abstract form features can be blended by combining the corresponding evaluation functions mathematically. However, in order to do this, the evaluation function which represents the abstract shape should be acquired, which is the opposite of the conventional process where the evaluation function is predefined. Accordingly, the method to convert the abstract shape to the evaluation function and operate it was developed on the basis of our previous study [4]. The method is composed of three steps. The outline of the system is illustrated in Fig. 9.2. In the first step, the evaluation function is acquired by the Evaluation Function Acquisition System (EFA). In the second, the acquired evaluation function is combined in the Evaluation Function Blending System (EFB). Finally, in the third step, a new abstract shape is created in the Blended Shape Creation System (BSC).

9.2.1 Evaluation Function Acquiring Process

In this method, an evaluation function is represented through a tree structure (Fig. 9.3 is an example of an evaluation function). This tree structure usually has two types of nodes: a terminal node and nonterminal node. Variable X_i, Y_i, or Z_i is

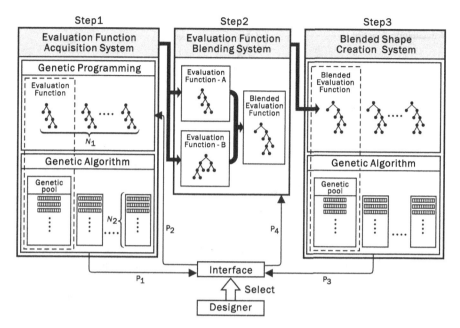

Fig. 9.2 System architecture for blending abstract shapes

Fig. 9.3 Example of an
evaluation function

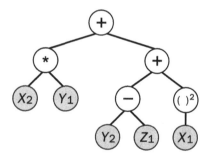

regarded a terminal node. A nonterminal node is represented by these operations: addition, subtraction, multiplication, and square.

EFA holds a fixed number (N_1) of functions as individuals in a Genetic Programming (GP) [2] and acquires the evaluation function by evolving these individuals. The individuals (evaluation functions) are the input to the Genetic Algorithms (GA) [1], and the abstract shapes which fit the evaluation functions are generated by using GA. Each evaluation function has the same number of genetic pools. Each genetic pool has a fixed number (N_2) of genes. GA generates the best-fitting abstract shape in the conventional shape representation method for each evaluation function. This method assumes that the abstract shapes are stored as images in the designer's mind. In Fig. 9.2, P_1 and P_2 denote the processes where the abstract shapes generated by the system are evaluated with respect to whether

Fig. 9.4 Crossing subtrees
of evaluation functions

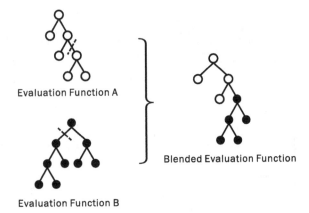

Evaluation Function A

Blended Evaluation Function

Evaluation Function B

or not they fit the images stored in the designer's mind through interaction with the designer. In P_1, the best-fitting abstract shapes for each evaluation function generated by GA are displayed on the computer screen. In P_2, the designer selects some abstract shapes which mostly fit the images stored in the designer's mind, and EFA identifies the evaluation functions corresponding to the selected abstract shapes, which will be used for the parents of GP. By repeating the above processes, EFA acquires the evaluation functions holding the abstract form features of the *base shapes*, which are imaged in the designer's mind.

9.2.2 Evaluation Function Blending Process

In the second step, EFB blends the evaluation functions acquired from EFA. In EFB, new evaluation functions are generated by crossing the subtrees of each evaluation function (Fig. 9.4). This method is similar to the notion of cross-breeding excellent parents to yield an excellent child by combining their genes via the method of Genetic Programming.

9.2.3 Blended Shape Creating Process

In the third step, the evaluation functions newly obtained from EFB are sent to BSC, and the shapes which fit these evaluation functions are generated by using a GA. In Fig. 9.2, P_3 and P_4 denote the processes where the generated blended shapes are evaluated through interaction with the designer. In P_3, the best-fitting abstract shapes generated by a GA for each of the newly generated evaluation functions are displayed on the computer screen. In P_4, the designer selects some abstract shapes which mostly fit the images which the designer wishes to generate,

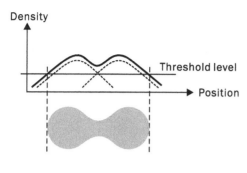
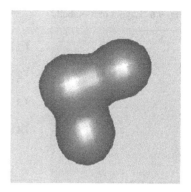

Fig. 9.5 Metaball method

and EFB identifies the newly generated evaluation functions corresponding to the selected abstract shapes, which will be used for the parents to be crossbred. By repeating P_3 and P_4, blended abstract shapes which match the designer's image are generated.

9.2.4 Abstract Shape Representation

The abstract shapes are represented on a computer by using the metaball method (Fig. 9.5) [3]. In the metaball method, each sphere has a certain density. If the summation of the metaballs' densities exceeds a threshold level, the area beyond the threshold level becomes the surface of the abstract shape. By changing the distance between metaballs and their densities, various shapes can be formed.

The metaballs are set on a lattice field. Only one metaball can exist in each lattice. In this method, the position of the metaballs is determined by using parameters which indicate the number of metaballs arranged in each row and column. For example, variables X_i indicate the number of metaballs arranged in column X_i (Fig. 9.6). By using these parameters, the position of the metaballs can be identified.

9.3 Experiment

We asked a subject to imagine objects such as "chair" and "pyramid" and synthesize them. In this experiment, each abstract shape was described by 12 variables (X_i, Y_i, and Z_i, where $i = 1, 2, 3, 4$). Then, the evaluation function was described as terminal nodes by the 12 variables and as nonterminal nodes by the following operations (Table 9.1).

Fig. 9.6 Layout of Metaballs

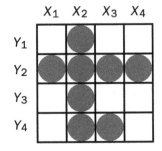

$X_1 = 1$		$Y_1 = 1$
$X_2 = 4$		$Y_2 = 4$
$X_3 = 2$		$Y_3 = 1$
$X_4 = 1$		$Y_4 = 2$

Table 9.1 Nonterminal nodes in evaluation functions

+	Addition
−	Subtraction
*	Multiplication
$(x)^2$	Square

Table 9.2 Parameters in GP

Population size (N_1)	120
Number of generations	30
Crossover rate	0.8
Mutation rate	0.6

Table 9.3 Parameters in GA

Population size (N_2)	50
Number of generations	1,000
Crossover rate	0.8
Mutation rate	0.05

In addition, the parameters for GP applied in EFA and those of the GA applied in EFA and BSC were given as follows (Tables 9.2 and 9.3). EFA and BSC displayed 12 shapes to the designer.

In all, two evaluation functions corresponding to the *base shapes* of "chair" and "pyramid" were acquired by using the two processes (P_1, P_2) mentioned above through EFA. Figure 9.7 illustrates the acquired shapes.

The acquired abstract shapes were blended into a new abstract shape through EFB and BSC. Figure 9.8 illustrates the newly created abstract shapes.

In Fig. 9.8, Shape 1 can be understood to be "pyramid which has the leg features of a chair", and Shape 2 can be understood to be "chair which has triangular legs".

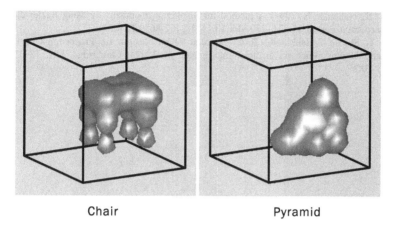

Chair   Pyramid

Fig. 9.7 Acquired abstract shapes through EFA

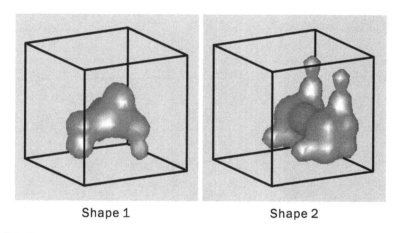

Shape 1   Shape 2

Fig. 9.8 Examples of blended abstract shapes

Although the created abstract shapes are too crude to be applied to the actual industrial design, the method introduced in this chapter is expected to be a fundament to establish a feasible method for synthesizing abstract form features.

References

1. Goldberg DE (2001) Genetic algorithms in search, optimization, and machine learning. Addison Wesley Longman Publishing Company, Boston
2. Koza JR (1993) Genetic programming: on the programming of computers by means of natural selection. The MIT Press, Cambridge

3. Nishita T, Nakamae E (1994) A method for displaying metaballs by using Bezier clipping. Comput Graph Forum 13:271–280. doi:10.1111/1467-8659.1330271
4. Taura T, Shiose S, Ishida R (2002) Strategic shape design. In: Proceedings of the 7th international conference on artificial intelligence in design. Cambridge, UK, 13–17 July, pp 371–382

Chapter 10
Synthesis of Motions: Practice of Concept Generation (2)

Abstract In this chapter, a method of synthesizing motions in order to generate a creative and emotional motion is developed. This method is based on the hypothesis that a creative motion, which is beyond the normal bounds of human imagination, can produce emotional impressions which *resonate* with the human mind. The proposed method figures in the *blending* and emphasizing of the features of motion on the space of the wavelet coefficients. In the experiment, the motions of "frog" and "snake" are blended and its impressions are evaluated. The results confirm the relevance of the hypothesis.

10.1 Creative and Emotional Motion

The significant ambition in design is to create emotional objects [5]. In this chapter, we attempt to establish a method of synthesizing motions in order to generate a creative and emotional motion which *resonates* with the deep-seated human feelings.

In recent years, with the availability of so many mediums of expression in today's information society, fields of design have widened their scope to include dynamic expression. In this chapter, we attempt to enhance the design of objects using motion. Humans have developed a great variety of motion behaviours, both for their own bodies and for products, ranging from vehicles to robots to animated figures. However, the conventional methods of generating these motion behaviours are usually based on actual or virtual figures created by the designer.

Compare this to another field of dynamic expression: music. In this field, instruments are created to facilitate the capacity of the creators to develop. It is arguable that musical instruments themselves trigger (or at least impact) the human feelings which people express via their music. We must note here that music differs from natural sound in that it is an artificial creation of humans, and

T. Taura and Y. Nagai, *Concept Generation for Design Creativity*,
DOI: 10.1007/978-1-4471-4081-8_10, © Springer-Verlag London 2013

we can be deeply impressed by music which evokes feelings beyond the normal bounds of human imagination, as mentioned in Chap. 2.

Humans receive emotional impressions not only from natural objects but from artefacts as well. We are deeply impressed by artefacts, such as pictures or music, just as we are impressed by nature, albeit in slightly different ways. (Here, the term 'emotional impression' suggests an active impression of the subject as something which touches the depths of the human mind, while the unadorned term 'impression' suggests a passive or static impression.)

We assume that 'creative' expression which is beyond the normal bounds of human imagination can produce such an emotional impression upon us (in this chapter, the term 'creative' is used to imply the notion of that which is beyond the normal bounds of human imagination). We develop a method to generate a 'creative' and 'emotional' motion by focusing on the causal relationship between a creative motion, which is beyond the normal bounds of human imagination, and emotional impressions, which *resonate* with the human mind. However, not just any motion which is beyond the normal bounds of human imagination is thought to inspire emotional impressions. In this method, we attempt to generate a creative motion by beginning with the motion of a natural object and transforming and elaborating on it.

To verify our hypothesis and the feasibility of our method, we constructed a computer system which is expected to trigger (or at least impact) the human mind, like musical instruments.

This method is developed by focusing on a motion behaviour which may be applied to something such as a moving logo. We anticipate that these kinds of moving corporate logos will soon become widespread on the Internet.

10.2 Method of Generating Creative and Emotional Motion

We still do not have a good instrument to describe and generate motions, unlike, for example, music. In addition, we can neither compose the motion, in contrast to music, which can be composed using music scores, nor play an instrument of motion in the same way that we play musical instruments.

To consider the issue of how we can generate motions, we can take some cues from dance. A dancer probably imagines the complete motion of a dance and composes the dance by moving his or her body. However, this method limits the imagination of motion to an individual's body. To overcome this limitation, dance directors use 'choreography', which is based on an expert knowledge of dance. Choreography implies 'dance-writing'—the art of generating sequences of movements, with each movement formed by various motions. Consider the famous choreographer Maurice Béjart, who created an advanced style of ballet [4]. He developed upon traditional ballet by integrating modern dance styles and adapting other dances to ballet. For instance, he studied Japanese Kabuki and blended the motions of Kabuki actors with those of Western ballet. By employing such methods, he partially blended many forms or motions that he borrowed from different sources, such as foreign dances, children's actions, and movie scenes. Béjart might not have been able to imagine all the motions nor generate novel motions without the use of the body.

Nonetheless, the human body greatly constrains the development of the generation of motions. If we develop a method and instruments that allow us to operate motions more freely, we can perhaps generate novel motions that are more creative and that go beyond the normal bounds of human imagination. In addition, we expect these novel motions to more strongly *resonate* with the human mind.

Nowadays, it is possible to create, record, and replay motions using computers. Thus, we can computationally generate new motions that humans cannot easily imagine. In this chapter, we discuss a method of synthesizing motions in order to generate a creative and emotional motion, by extending the principles set forth in our previous works [7, 8].

10.2.1 A Mimic of Natural Objects

Humans have evolved within the natural environment and are thought to have images of nature imprinted in their minds. Indeed, humans have created many artefacts which are based on or suggested by natural objects, and their motions are both unique and charming [1]. For example, research in the field of biologically inspired design has pointed out motions developed through the application of biological phenomena [9]. It is for this reason that we use natural objects as a source for generating a creative and emotional motion which extends beyond the normal bounds of human imagination and *resonates* with the human mind.

10.2.2 Blending of Motions

A motion which is generated solely by mimicking natural objects cannot be said to extend beyond the normal bounds of human imagination. On the other hand, *concept blending* in *concept synthesis* leads to a highly *innovative design idea*, as mentioned in Chaps. 4 and 5. We apply the *concept blending* to the process of generating a creative and emotional motion, and we develop a method of blending the motions mimicking natural objects in order to generate a more creative motion.

10.2.3 Emphasis on Rhythmic Features

In order to generate a more creative motion which extends far beyond the normal bounds of human imagination, we attempt to emphasize the rhythmic features of

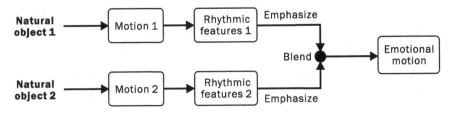

Fig. 10.1 Outline of the Rhythmic Feature–based Motion Blending method

motion, that is, the changes in the number of angles of joints and in the angular velocities. Rhythm in music involves the interrelationship between the accented (strong) and the unaccented (weak) beat [2]. Incidentally, accents which are produced by stress (dynamics) imply the dynamic intensification of a beat, that is, an emphasis implied through the use of a louder sound. For example, in Western music literature, p (piano) implies 'soft', while f (forte) implies 'loud'. On the basis of these considerations in the field of music, we attempt to emphasize the rhythmic features of a motion by increasing or reducing its amplitude. Using this method, we anticipate that motions which extend far beyond the normal bounds of human imagination can be generated.

10.2.4 Rhythmic Feature–Based Motion Blending

On the basis of the considerations mentioned above, we developed a method of generating a creative and emotional motion which incorporates the blending of motions mimicking natural objects and an emphasis on rhythmic features (hereafter, referred to as Rhythmic Feature–based Motion Blending (RFM Blending)). The outline of this method is illustrated in Fig. 10.1.

The method used for the calculation of RFM Blending is given below. First, we obtain the motions of natural objects as a source for a creative and emotional motion (hereafter, referred to as *base motions*). The rhythmic features are extracted by conducting a frequency analysis of each obtained motion. In this analysis, we decided to use wavelet analysis because it can process both the phasing and frequential characteristics in the same operation [3]. The wavelet coefficients obtained from wavelet analysis are then used as the rhythmic features of motion. Let $\omega_1(t)$ be the tth rhythmic feature (wavelet coefficient) of natural object 1. The emphasized rhythmic feature is then defined by the following Eq. (10.1).

$$\omega_1(t) = \begin{cases} m_1\omega_1(t) & (p_1 \le |\omega_1(t)|) \\ \omega_1(t) & (q_1 < |\omega_1(t)| < p_1) \\ n_1\omega_1(t) & (|\omega_1(t)| \le q_1) \end{cases} \qquad (10.1)$$

Here, p_1 and q_1 are the thresholds for emphasis of a rhythmic feature, with $m_1 \geq 1$ and $n_1 \leq 1$. This emphasis makes a large motion larger and a small motion smaller. The rhythmic feature of the motion of another natural object is emphasized in the same way. The two rhythmic features are then blended together. The operation of blending rhythmic features is defined in Eq. (10.2). Here, R_1 and R_2 are the rhythmic features of the motion of natural object 1 and natural object 2, respectively, and C_1 and C_2 are the weights for the rhythmic features of natural objects 1 and 2, respectively.

$$R_{blending}(\omega(t)) = C_1 R_1(\omega_1(t)) + C_2 R_2(\omega_2(t)) \tag{10.2}$$

Inverse wavelet transform is performed on the blended rhythmic features so that a new motion is generated.

10.3 Procedure to Generate Creative and Emotional Motion

On the basis of the method described above, a computer system for generating a creative and emotional motion was developed. The procedure to implement the system is summarized in the following steps.

- Step 1: Obtain the angle $\theta(t)$ in the sequential order by selecting each of the four joints of a natural object as a characteristic point.
- Step 2: Calculate the angular velocity $\dot{\theta}(t)$ of each joint from the changes in its angles.
- Step 3: Perform wavelet analysis for the angular velocity calculated in Step 2. In this method, we use Daubechies 8 wavelets as wavelet prototype functions, since these wavelets are widely used. The wavelet coefficients obtained in Step 3 are used as the rhythmic features.
- Step 4: The rhythmic features are emphasized using Eq. (10.1). In the process of emphasis, designers can decide p, q, m, and n according to their individual criteria.

Steps 1–4 are performed on each joint of the natural object.

- Step 5: The rhythmic features of two natural objects, which were emphasized in Step 4, are blended using Eq. (10.2). Here, too, motion designers can decide the weights for each rhythmic feature according to their individual criteria.
- Step 6: Inverse wavelet transform is performed on the blended rhythmic features of each joint $R_{blending}(\omega(t))$, and angular velocities $\dot{\theta}_{blending}(t)$ are obtained. The angle $\theta_{blending}(t)$ of each joint is calculated from the angular velocities.
- Step 7: The motion is created by transforming the angle of each joint to that of the design object.

We developed a computer system which could perform the steps described above.

Table 10.1 Types of motions

	Motion 1	Motion 2	Motion 3
p	Top 25%	–	
m	4.0		
q	–	Bottom 25%	
n		0.25	

10.4 Experiment

An experiment was conducted to investigate the effect of extending a motion beyond the normal bounds of human imagination, by emphasizing the rhythmic features of the motion using RFM Blending and confirm the feasibility of the method. In this experiment, three types of motion were created which differed in the way they emphasized rhythmic features.

10.4.1 Generation of Creative and Emotional Motion Using RFM Blending

The selected characteristic motions of "frog" and "snake" were used as the *base motions*, since they both have unique ways of moving which people tend to recognize easily. We chose the arms of a virtual robot in CG as a design object. In order to generate a motion which would be beyond the normal bounds of human imagination, we set no limitations on the disposition of the robot's arms; for example, neither limiting the joint angle nor preventing a collision. The types of rhythmic features emphasized are listed in Table 10.1. Here, the determinations of p and q were calculated beforehand from the average of each rhythmic feature of each joint in Step 3. The weights for each rhythmic feature C_1: C_2 were determined as 1:1. Motion 1 was expected to enhance a large motion by making it larger, while Motion 2 was expected to diminish a small motion by making it smaller. Motion 3 emphasized no rhythmic features.

The procedure is illustrated in Fig. 10.2. Steps 1–7 are performed to generate the motion for the experiment. The examples of newly generated motion are shown in Fig. 10.3, which illustrates these motions in a time sequence with 2-second intervals.

10.4.2 Evaluation of Generated Motions

The Semantic Differential (SD) method [6] was used to evaluate the motions. In the SD method, subjects are asked to evaluate their impressions by identifying the value of the predefined scale on the answer form. In this experiment, the motions

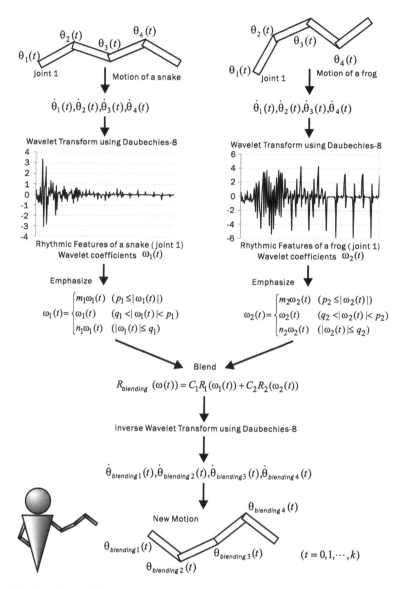

Fig. 10.2 Procedure of blending motions

were evaluated according to 10 terms on a seven-point scale. The terms are presented in Fig. 10.4. A total of 12 subjects participated in the evaluation. After being shown three motions, they were asked to evaluate the motions. The subjects were undergraduates and graduate students. In order to control the sequence effect of the order of presentation, six of the subjects evaluated Motion 1 first, Motion

Motion 1   Motion 2   Motion 3

Fig. 10.3 Example of generated motion (although Motions 2 and 3 appear very similar here, they look different in an actual motion)

3 second, and Motion 2 last, while the others evaluated Motion 2 first, Motion 3 second, and Motion 1 last.

The SD profile obtained from the experiment is illustrated in Fig. 10.4. The profile of the two groups did not show a significant differences ($F(1, 10) = 1.68$, n.s.). The points in the figure show the average of all subjects for each term for each motion. In the figure, we can see that the following five terms elicited a significantly different response to Motion 1 (a large motion enhanced to become larger) versus Motion 3 (no feature emphasized): term 5, 'Easy to mimic the body—Difficult to mimic' ($F(2, 200) = 4.74, p < 0.01$); term 6, 'Compelling—Not compelling' ($F(2, 200) = 10.53, p < 0.01$); term 7, 'Vivid—Vapid' ($F(2, 200) = 4.74, p < 0.01$); term 8, 'Complicated—Simple' ($F(2, 200) = 12.52, p < 0.01$); and term 9, 'Dynamic—Static' ($F(2, 200) = 4.07, p < 0.01$). Furthermore, the following three terms elicited a significantly different response to Motion 1 versus Motion 2 (a small motion diminished to become smaller): term 6, 'Compelling—Not compelling' ($F(2, 200) = 10.53, p < 0.01$); term 7, 'Vivid—Vapid' ($F(2, 200) = 4.74$,

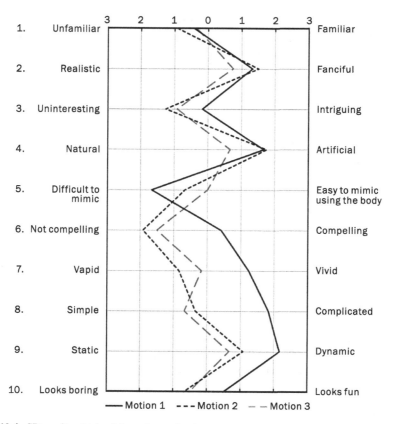

Fig. 10.4 SD profile obtained from the evaluation

$p < 0.01$); and term 8, 'Complicated—Simple' ($F(2, 200) = 4.74$, $p < 0.01$). On the other hand, the values of Motions 1 and 2 were close to one another: term 2, 'Fanciful—Realistic', and term 4, 'Artificial—Natural', while both these values were different from the value obtained for Motion 3.

Accordingly, a significant difference can be found between the motions for terms 5, 6, 7, 8, and 9. Among these terms, term 5, 'Easy to mimic with the body—Difficult to mimic' and term 8, 'Complicated—Simple', are considered to be related to creative expression which extends beyond the normal bounds of human imagination. On the other hand, term 6, 'Compelling—Not compelling'; term 7, 'Vivid—Vapid'; and term 9, 'Dynamic—Static' are considered related to an emotional impression which *resonates* with the human mind. These results demonstrate that there is a tendency for motions which extend beyond the normal bounds of human imagination to produce emotional impressions, thus confirming our hypothesis. However, not just any motion which extends beyond the normal bounds of human imagination can produce emotional impressions. The method developed in this chapter involves two strategies which may at first glance seem contradictory: one is to use natural objects as a source and the other is to elaborate on them, so that they extend beyond the normal bounds of human imagination. We believe that these seemingly contradictory strategies are effective for generating a truly creative and emotional motion.

References

1. Chakrabarti A, Sarkar P, Leelavathamma B, Nataraju BS (2005) A functional representation for aiding biomimetic and artificial inspiration of new ideas. AI EDAM 19:113–132. doi:10.1017/S0890060405050109
2. Cooper GW, Meyer LB (1960) The rhythmic structure of music. University of Chicago Press, Chicago
3. Daubechies I (1992) Ten lectures on wavelets. Society for Industrial Mathematics, Philadelphia
4. Krasovskaya V (1990) An anatomy of genius. Danc Chron 13:82–88. doi:10.1080/01472529008569026
5. Norman DA (2003) Emotional design: why we love (or hate) everyday things. Basic Books, New York
6. Osgood CE, Suci GJ, Tannenbaum PH (1957) The measurement of meaning. University of Illinois Press, Urbana
7. Taura T, Nagai Y (2010) Designing of emotional and creative motion. In: Fukuda S (ed) Emotional engineering: service development. Springer, London
8. Tsujimoto K, Miura S, Tsumaya A, Nagai Y, Chakrabarti A, Taura T (2008) A method for creative behavioral design based on analogy and blending from natural things. In: Proceedings of ASME 2008 international design engineering technical conference and computers and information in engineering conference. Brooklyn, New York, 3–6 August (CD-ROM)
9. Vattam S, Helms M, Goel AK (2007) Biologically inspired innovation in engineering design: a cognitive study. Graphics, visualization and usability center Technical report, Georgia Institute of Technology, GIT-GVU-07-07

Chapter 11
Synthesis of Functions: Practice of Concept Generation (3)

Abstract In this chapter, a method of synthesizing functions in order to support the design of a new function is developed. The method of deriving a new lower level function structure is systematized from a linguistic viewpoint. The method classifies the function synthesis processes into two types: 'integration operation', which corresponds to *concept blending*, and 'conversion operation', which corresponds to *property mapping*. Furthermore, a thesaurus is developed to support the linguistic operations. This method is expected to be a fundament of the feasible framework for the 'design of function', which may help us answer the question: 'What should we create?'

11.1 Design of Function

Currently, our living environment is filled with various products, and truly excellent products have become a greater requirement of consumers than ever before. In this situation, 'What should we create?' has become an important topic of discussion, whereas in the past, it was 'How should we realize the product?' In the context of engineering design, we identify 'What' as a function. Accordingly, in the very early stage of engineering design, the focus should be on not only deriving a novel mechanism but also creating a novel function.

So far, function description has been used to represent the required specifications, and various function-description-based models of the design process have been developed, aimed at design support [1–3, 7]. However, because the aim of these methods was to answer the question: 'How should we realize the given function?' it might be difficult to answer the question: 'What should we create?'

It is necessary to investigate the functions from the viewpoint of *concept generation*. In this chapter, we discuss the issue of 'generating a new function', and propose an associated methodology. Indeed, by focusing on *concept blending*

T. Taura and Y. Nagai, *Concept Generation for Design Creativity*,
DOI: 10.1007/978-1-4471-4081-8_11, © Springer-Verlag London 2013

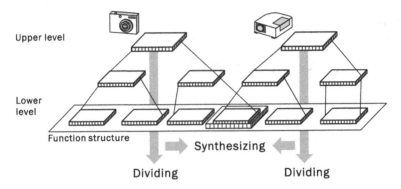

Fig. 11.1 Synthesis of the function structure

and *property mapping*, a method which generates a new function from some functions (hereafter, referred to as **base functions**) is developed. Specifically, we derive a new lower level function structure (hereafter, the lower level functions structure is referred to as 'function structure'), which is original and feasible (Fig. 11.1).

We focus on the lower level function structures because the new function becomes meaningful only when the function can be implemented into a concrete mechanism and dividing the upper level function into a lower level function structure makes the implementation practical. This is because the generation of a function is different from idea generation, which does not require the implementation of a mechanism. To support this function generation in a feasible manner, it is necessary to elaborate on current functional decomposition processes in the field of engineering design and develop a method for mapping between upper and lower level functions. In our previous study, a method was developed which could be implemented on a computer, supporting the functional decomposition process. This method identifies the hierarchical relations between upper and lower level functions in the functional decomposition process, from the viewpoint of linguistic hierarchical relations between words describing the upper and lower level functions [8] (Fig. 11.2).

Furthermore, we developed a thesaurus which supports the mapping between the upper and lower level functions. In this chapter, we expand this method and propose another which can support the generation of a new function structure by synthesizing *base functions*. For this, we first conducted a case study and determined how to synthesize *base functions*. Accordingly, we identified operating types for synthesizing functions. Next, we formulated the identified operating types and simultaneously developed a thesaurus which could support the procedure of synthesizing functions. Finally, we developed a computer system which implemented the function synthesizing process.

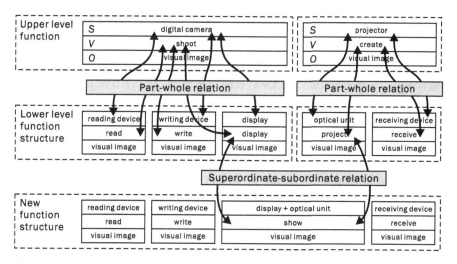

Fig. 11.2 Linguistic hierarchy in the function hierarchy

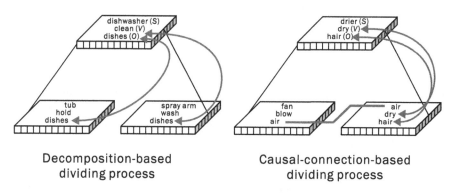

Decomposition-based
dividing process

Causal-connection-based
dividing process

Fig. 11.3 Types of dividing processes (FDP)

11.2 Function Synthesis

We developed a method of obtaining lower level functions from upper level functions (hereafter, referred to as Function Dividing Process (FDP)). In the study, we conducted a case study and found the types in the FDP. There exist two types of divisions: the 'decomposition-based dividing process' and the 'causal-connection-based dividing process', as illustrated in Fig. 11.3.

The causal-connection-based dividing process deals with the causal effects of physical phenomena among the components (see the figure on the right side of Fig. 11.3). It is said that two functions are causally connected when an object of one function is equal to a subject of the adjacent function at the same level. In the causal-connection-based dividing process, a chain of causal relations connects the lower level functions; an object of the terminal lower level function is equal to that of the upper level function. In addition, the verb of the terminal lower level function is equal to that of the upper level function. In the decomposition-based dividing process, the subset of the upper level function is the entity which realizes the subset of subjects from the lower level functions; the verb of the upper level function is the action which realizes the subset of verbs from the lower level functions. The object of each lower level function is equal to that of each upper level function (see the figure on the left side of Fig. 11.3).

Here, a function is defined as 'an entity's behaviour which plays a special role' and is described using a subject (S), verb (V), and object (O). On the basis of the above considerations, we develop a method to generate a new function structure by synthesizing some existing functions (hereafter, referred to as 'function synthesis in the FDP').

11.2.1 Definition of the Function Synthesis in the FDP

We define the function synthesis in the FDP as follows:

$$m : F_{upper} \rightarrow F_{lower}, \tag{11.1}$$

where F_{upper} denotes the set of upper level functions and F_{lower} denotes the set of new lower level functions. Figure 11.4 illustrates the flow of the process of generating a new function. In this figure, BF_{lower} denotes the set of lower level functions, which are obtained by dividing the upper level functions. First, a new lower level function structure is obtained through the function synthesis process (from the upper to the middle figure in Fig. 11.4). Next, a new upper level function is obtained from the newly obtained lower level function structure (from the middle to the lower figure in Fig. 11.4). In this chapter, we discuss the process used to obtain a new lower level function structure via the function synthesis process.

11.2.2 Classification of the Function Synthesis in the FDP

We conducted a case study to formulate the process of the function synthesis in the FDP. We analysed 13 examples of engineering products, which can be viewed as the result of synthesizing the functions of two existing products. The case study results revealed that the function synthesis processes can be classified into two types: 'integration operation', which corresponds to *concept blending*, and 'conversion operation', which corresponds to *property mapping*. The integration operation blends two functions. For example, by blending the lower level function of "digital camera"—'display displays the visual image'—and the lower level

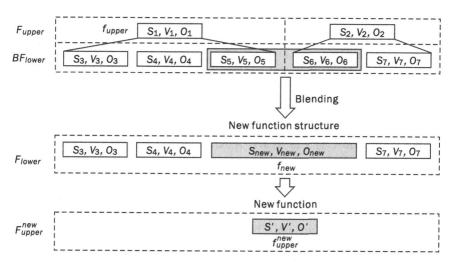

Fig. 11.4 Process of generating a new function

function of "projector"—'optical unit projects the visual image'—into a new lower level function '{display and optical unit} {displays and projects} the visual image', a new engineering product—"digital camera with projector"—can be generated. On the other hand, the conversion operation replaces an existing function with another function. For example, by replacing the function of 'battery generates electricity' in "flashlight" with the function of 'hand generator generates electricity' in "hand generator", the function of "flashlight with a hand generator" can be generated. We describe the formulation of these operations in more detail later.

On the basis of these analyses, the function synthesis in the FDPs can be classified into 'function synthesis in the FDP by integration' and 'function synthesis in the FDP by conversion'.

In this method, we describe a function, $f = (S, V, O)$, by using a subject (S), verb (V), and object (O). The subject (S) is assumed to represent a mechanism to achieve a verb (V) and an object (O). To capture a function hierarchy, Pahl and Beitz [6] propose a function structure diagram in which functions are regarded as input/output relations represented by nouns and verbs. Miles [4] proposes a method of defining a function in terms of a linguistic expression for value engineering. In these methods, verbs and objects constitute a type of linguistic expression. By taking these considerations into account, a function is understood to be constituted by 'a subject' and 'a verb and an object'. Accordingly, the function synthesis processes are classified into 'to obtain a new subject' and 'to obtain a new verb or a new object, or both'. Table 11.1 presents the classification of the function synthesis.

Table 11.1 Classification of the function synthesis process

	Integration	Conversion
New subject	Function synthesis in the FDP by subject integration	Function synthesis in the FDP by subject conversion
New verb and/ or object	Function synthesis in the FDP by verb or object integration	Function synthesis in the FDP by verb and object conversion

11.2.3 Definition of the Word Abstraction–Concretion Operation

To conduct the integration and conversion operations, it is necessary to determine whether two functions can be equated. Therefore, we introduced an operation which abstracts or concretizes the words described in the functions, and call it the 'word abstraction-concretion operation'. The abstraction operation obtains an abstract word from a given word by extracting its common and/or essential nature. An example is to obtain 'food' from 'edible fruit'. The concretion operation obtains a more concrete word from the given word by clearly indicating the content of the given word. An example is to obtain 'apple' from 'edible fruit'. We define a set of words, $P(w)$, obtained by the abstraction-concretion operation for a word w, as follows.

$$P(w) = \{w_i | w_i \text{ is a word obtained by abstracting or concreting } w\} \quad (11.2)$$

Here, suppose w_1 is included in $P(w_2)$; that is, $w_1 \in P(w_2)$. In this case, when both w_1 and w_2 are nouns, we define the relation between them as a 'superordinate-subordinate relation of nouns'. Similarly, when both w_1 and w_2 are verbs, we define the relation between them as a 'superordinate-subordinate relation of verbs'. These relations are well-known as hierarchical semantic relations. Therefore, we can identify them in existing dictionaries and thesauri.

11.2.4 Definition of the Word Match

We define the situation in which two words are found in the word abstraction–concretion operation as a 'word match'. A set, $I(w)$, of words which match another word w is defined as follows.

$$I(w) = \{w_i | w_i = w \vee w_i \in P(w) \vee 0 < |P(w) \cap P(w_i)|\} \quad (11.3)$$

A set, $M(w_1, w_2)$, of words where the two words match each other is defined as follows.

$$M(w_1, w_2) = \{w_i | w_i = w_1 \in P(w_2) \vee w_i = w_2 \in P(w_1) \vee w_i \in P(w_1) \cap P(w_2)$$
$$\vee w_i = w_1 = w_2\}$$

$$(11.4)$$

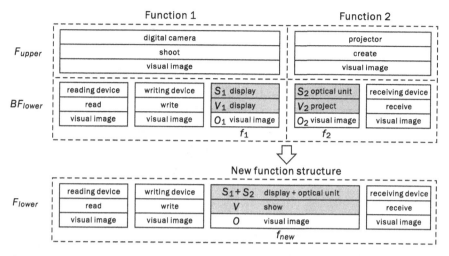

Fig. 11.5 Example of function synthesis in the FDP by subject integration

11.2.5 Formulation of the Function Synthesis in the FDP

By introducing the word abstraction-concretion operation and the word match, the equatability among verbs or nouns can be determined. In this section, by using these operations, we formulate the function synthesis in the FDP, which is implemented by integration or conversion.

11.2.5.1 Function Synthesis in the FDP by Integration

As listed in Table 11.1, integration operations can be classified into two operating types. When a word match exists between the verbs of two functions as well as the objects, we define the operation of blending the two functions by joining their subjects as 'function synthesis in the FDP by subject integration'. When a word match exists between the subjects of two functions as well as between the verbs or objects of the two functions, we define the operation of blending the two functions by joining their verbs or objects as 'function synthesis in the FDP by verb or object integration'. We describe the operation of joining two words using the '+' operator. Using this '+' operator, we formulate these operations as follows.

(1) Function synthesis in the FDP by subject integration (Fig. 11.5)

$$m_1 : F_{\text{upper}} \rightarrow F_{\text{lower}}$$
$$f_1 \in BF_{\text{lower}}, \ f_1 = (S_1, V_1, O_1), \ f_2 \in BF_{\text{lower}}, \ f_2 = (S_2, V_2, O_2)$$
$$\text{when } V \in M(V_1, V_2), \ O \in M(O_1, O_2)$$
$$f_{\text{new}} = (S_1 + S_2, V, O), \ F_{\text{lower}} = \{f_i | f_i \in (BF_{\text{lower}} - \{f_1\} - \{f_2\}) \cup \{f_{\text{new}}\}\}$$

$$(11.5)$$

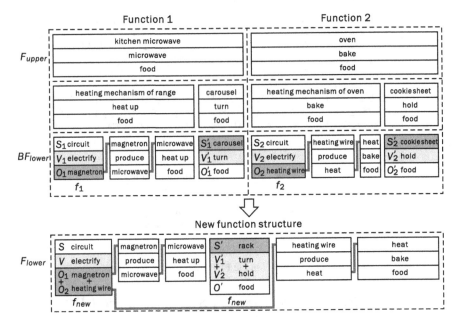

Fig. 11.6 Example of function synthesis in the FDP by verb or object integration

(2) Function synthesis in the FDP by verb or object integration (Fig. 11.6)

$$m_2 : F_{\text{upper}} \rightarrow F_{\text{lower}}$$

$$f_1 \in BF_{\text{lower}}, \ f_1 = (S_1, V_1, O_1), \ f_2 \in BF_{\text{lower}}, \ f_2 = (S_2, V_2, O_2)$$

$$\text{when } S \in M(S_1, S_2), \ V \in M(V_1, V_2) \text{ or } O \in M(O_1, O_2) \qquad (11.6)$$

$$f_{\text{new}} = (S, V_1 + V_2, O) \vee (S, V, O_1 + O_2)$$

$$F_{\text{lower}} = \{f_i | f_i \in (BF_{\text{lower}} - \{f_1\} - \{f_2\}) \cup \{f_{\text{new}}\}\}$$

11.2.5.2 Function Synthesis in the FDP by Conversion

Similarly, conversion operations can be classified into two operating types. We define the operation of replacing a function with another function when a word match exists between the verbs and objects of the two functions, as a 'subject conversion operation'. An example of this conversion operation is the new concept of "flashlight with hand generator", discussed in Sect. 11.2.2. We formulate 'function synthesis in the FDP by subject conversion', as follows.

(3) Function synthesis in the FDP by subject conversion (Fig. 11.7)

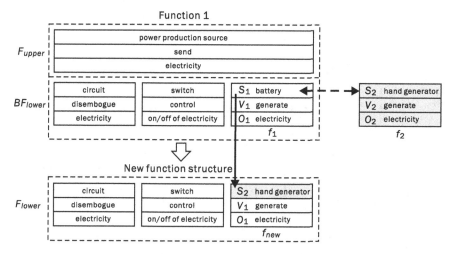

Fig. 11.7 Example of function synthesis in the FDP by subject conversion

$$m_3 : F_{\text{upper}} \rightarrow F_{\text{lower}}$$
$$f_1 \in BF_{\text{lower}}, \ f_1 = (S_1, V_1, O_1), \ f_2 = (S_2, V_2, O_2)$$
$$\text{when } V_1 \in I(V_2), \ O_2 \in I(O_2) \tag{11.7}$$
$$f_{\text{new}} = (S_2, V_1, O_1), \ F_{\text{lower}} = \{f_i | f_i \in (BF_{\text{lower}} - \{f_1\}) \cup \{f_{\text{new}}\}\}$$

On the other hand, the verb and object conversion operation can be viewed as an operation to capture the role played by a subject from a different perspective. For example, the function of 'fan blows air' can be captured instead as a function of 'fan cools something', simply by changing one's viewpoint. Although engineering products manifest several functions satisfying the requirements originally provided by a designer, some other functions can be observed under different circumstances. For example, chairs are designed to provide a seat for a human; however, they may also function as stepladders. A function satisfying the requirements provided by a designer can be defined as a 'visible function', whereas different functions which manifest under different circumstances are defined as 'latent functions' [9]. A latent function can be understood as another function which is discovered when the product is observed from a different viewpoint. From this understanding, a method of reasoning out latent functions based on some visible function was proposed [5]. The method entails inferring a latent function by transferring a function from another engineering product to the visible function of the focal engineering product. Concretely, when there is a *commonality* between the lower level functions of these two products, a latent function—$f_1^{\text{new}} = (S_1, V_2, O_2)$—can be obtained by replacing verb V_1 and object O_1 from the function—$f_1 = (S_1, V_1, O_1)$—of the product, with verb V_2 and object

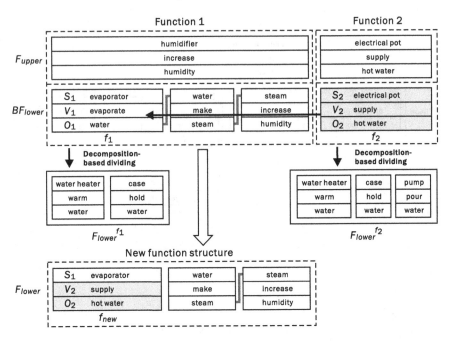

Fig. 11.8 Example of function synthesis in the FDP by verb and object conversion in decomposition-based dividing (when a causal connection is not included in both F_{lower}^{f1} and F_{lower}^{f2})

O_2 of a function—$f_2 = (S_2, V_2, O_2)$—of another product. This operation is nothing more than the verb and object conversion operation.

On the basis of the above considerations, we formulate a 'function synthesis in the FDP by verb and object conversion'. We formulate this process to address two cases: when a causal connection is included in both the lower functions, and when it is not.

(4-1) Function synthesis in the FDP by verb and object conversion, when a causal connection is not included in both F_{lower}^{f1} and F_{lower}^{f2} (Fig. 11.8). This case is valid only when there are more than N common functions between F_{lower}^{f1} and F_{lower}^{f2} (i.e. $|F_{\text{lower}}^{f1} \cap F_{\text{lower}}^{f2}| \geq N$ (constant)).

$$m_4 : F_{\text{upper}} \rightarrow F_{\text{lower}}$$
$$f_1 \in BF_{\text{lower}}, \; f_1 = (S_1, V_1, O_1), \; f_2 = (S_2, V_2, O_2)$$
$$\text{when } |F_{\text{lower}}^{f1} \cap F_{\text{lower}}^{f2}| \geq N(\text{constant}) \tag{11.8}$$
$$f_{\text{new}} = (S_1, V_2, O_2), \; F_{\text{lower}} = \{f_i | f_i \in (BF_{\text{lower}} - \{f_1\}) \cup \{f_{\text{new}}\}\}$$

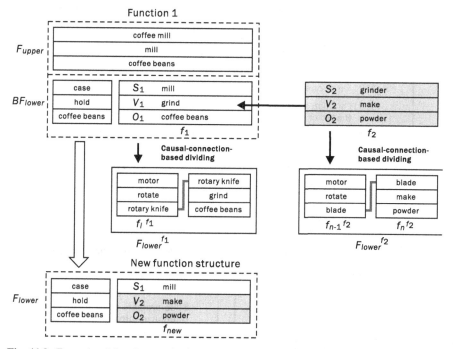

Fig. 11.9 Example of function synthesis in the FDP by verb and object conversion in causal-connection-based dividing (when a causal connection is included in both F_{lower}^{f1} and F_{lower}^{f2})

Here, f_1 denotes a lower level function of some upper function and f_2 denotes a lower level function of some other upper function. In addition, F_{lower}^{f1} denotes the set of lower level functions when the upper level function is f_1, and F_{lower}^{f2} denotes the set of lower level functions when the upper level function is f_2.

(4-2) Function synthesis in the FDP by verb and object conversion, when a causal connection is included in both F_{lower}^{f2} and F_{lower}^{f2} (Fig. 11.9)

$$m_5 : F_{upper} \rightarrow F_{lower}$$
$$f_1 \in BF_{lower}, f_1 = (S_1, V_1, O_1), f_2 = (S_2, V_2, O_2)$$
$$|F_{lower}^{f2}| = n,$$
$$f_i^{f1} \in F_{lower}^{f1}, f_{n-1}^{f2} \in F_{lower}^{f2}, f_n^{f2} \in F_{lower}^{f2}, \qquad (11.9)$$
$$f_i^{f1} = (\bar{S}_i, \bar{V}_i, \bar{O}_i), f_{n-1}^{f2} = (\overline{S_{n-1}}, \overline{V_{n-1}}, \overline{O_{n-1}}), f_n^{f2} = (\overline{S_n}, \overline{V_n}, \overline{O_n})$$
$$\text{when } \overline{O_{n-1}} = \overline{S_n}, \overline{V_n} = V_2, \overline{O_n} = O_2, \bar{S}_i = \overline{S_{n-1}}, \bar{V}_i = \overline{V_{n-1}}$$
$$f_{new} = (S_1, V_2, O_2), F_{lower} = \{f_i | f_i \in (BF_{lower} - \{f_1\}) \cup \{f_{new}\}\}$$

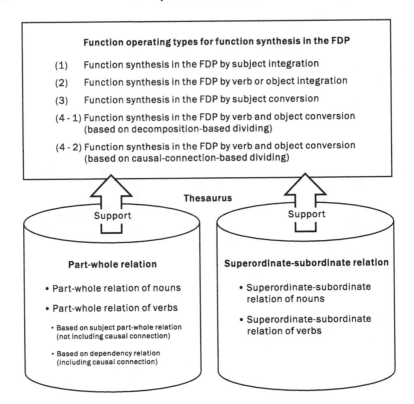

Fig. 11.10 Relationship between function operating types for function synthesis in the FDP and thesaurus

11.3 Thesaurus

Our proposed method is significant in that it facilitates identification of the hierarchical relations between the upper and lower level functions in the FDP, on the basis of the linguistic hierarchical relations between the words in the upper level functions and those in the lower level functions. Figure 11.10 shows the word relationships which compose the thesaurus supporting function synthesis in the FDP.

The inputs to the thesaurus are a word (a noun or a verb) and a type of word relationship (a superordinate-subordinate relation or a part-whole relation). The outputs are pairs of words—the input word and another word, which is determined by the input relation to the input word. For example, when 'product name' and 'noun part-whole relation' are provided as input, the product name and its components are provided as an output pair.

The thesaurus was constructed in the following manner.

For the superordinate-subordinate relation, we used the relations which are contained in the existing concept dictionaries, since a sufficient number of relations are already contained in these concept dictionaries.

Table 11.2 Examples of the part-whole relation of verbs based on a subject part-whole relation

Subject part-whole relation (part:whole)	Verb part-whole relation (part:whole)
fan:vacuum cleaner	blow:suck up
motor:hair clipper	drive:cut
sensor:rice cooker	measure:cook
burner:cooking stove	burn:bake

Table 11.3 Examples of the dependency relation between subject and verb

subject–verb
burner–burn, cam–rotate, filter–deodorize, heater–generate, rotor–drive, refrigerator–store, mirror–turn, holding member–hold, regulator–provide

Table 11.4 Examples of the dependency relation between verb and object

verb–object
cut–hair, cool–engine, drive–blade, increase–pressure, pump–water, rotate–gear, rotate–impeller, turn–wheel

Table 11.5 Examples of the part-whole relation of verbs based on causal connection

{(upper level function S, V:premise S, V, O), (upper level function S, V:consequence S, V, O)}
{(fan, produce:motor, rotate, impeller),   (fan, produce:impeller, produce, air)}   {(transmission, change:control valve, actuate, clutch),   (transmission, change:clutch, change, gear ratio)}

Regarding the part-whole relation, we need to develop a method for extracting the relations, since they are not sufficiently addressed in most of these dictionaries.

First, we attempted to extract 'part-whole relations of verbs' in the case where a causal connection between lower level functions is not included [8]. In this attempt, we extracted the 'part-whole relations of verbs', on the basis of the part-whole relation of subjects, from documents. Table 11.2 presents examples of the obtained results. When the relation between 'vacuum cleaner' and 'fan' is a part-whole relation between the subjects, 'suck up', which is an action of 'vacuum cleaner', and 'blow', which is an action of 'fan', are determined to have a verb part-whole relation.

Next, we attempted to extract part-whole relations for a verb in the case where a causal connection between lower level functions is included. The method for the extraction of the relations entails the following steps.

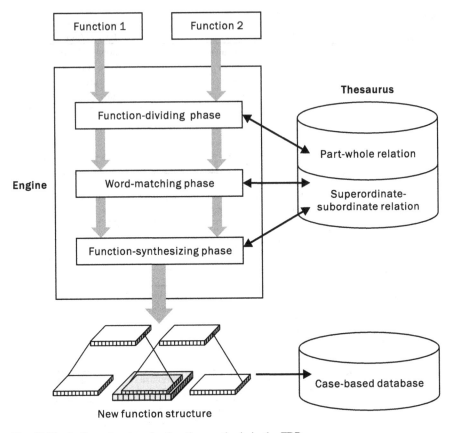

Fig. 11.11 Outline of system for function synthesis in the FDP

- Step 1: Extract the dependency relation between the subject and the verb from a set of documents.
- Step 2: Extract the dependency relation between the verb and the object from a set of documents.
- Step 3: Derive the part-whole relation of the verb, on the basis of causal connection, by incorporating the dependency relations between the nouns and the verbs (subject-verb and verb-object) into the part-whole relation of the verb, which is determined from the part-whole relation of the subject.

We extracted the dependency relations between subjects and verbs, and the dependency relations between verbs and objects, from patent abstracts in Japan and service manual texts (a total of approximately 8,000 lines). As a result, we extracted 11,198 dependency relations between subjects and verbs, and 13,348 dependency relations between verbs and objects. Table 11.3 presents examples of the extracted subject-verb dependency relations. Table 11.4 presents examples of the extracted

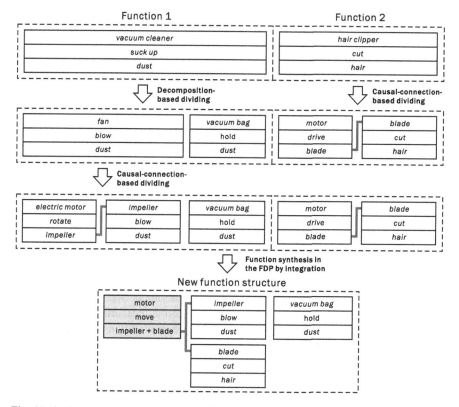

Fig. 11.12 Generated function structure of hair clipper with vacuum cleaner

verb-object dependency relations. Table 11.5 presents examples of the part-whole relation of verbs on the basis of causal connection. For example, for the upper level function where the subject is 'fan' and the verb is 'produce', lower level functions (e.g. 'motor, rotate, impeller' and 'impeller, produce, air') are causally connected, wherein a part-whole relation of verbs, based on a part-whole relation of subjects, is valid between the upper level function and each lower level function.

11.4 System for Function Synthesis in the FDP

A system for function synthesis in the FDP was composed of an input interface, engine for the function synthesis in the FDP, output interface, thesaurus, and case-based database. Figure 11.11 illustrates an outline of the system for the function synthesis in the FDP. We assumed a usage scenario wherein a designer is synthesizing two engineering products. The designer inputs two sets of subjects, verbs, and objects, in the form of (S, V, O), for the two engineering products. Then, two input functions are synthesized via the function-dividing phase, word-matching phase, and function-synthesizing phase, and finally, a new function structure is shown on the computer screen. The thesaurus of part-whole relations supported the function-dividing phase. The thesaurus of superordinate-subordinate relations supported the word-matching and function-synthesizing phases. The function structure obtained by the above process using the system was stored in the case-based database.

11.5 A Trial to Generate a New Function Structure

To confirm the feasibility of the proposed method for the function synthesis in the FDP, we conducted a trial to generate a new function structure of "hair clipper with a vacuum cleaner" by blending the functions of "vacuum cleaner" and "hair clipper". We conducted the trial as naturally as possible, by not basing the operations only on the thesaurus when the appropriate words were not found in it. Figure 11.12 shows the function structure constructed for "hair clipper with the vacuum cleaner". Verbs and nouns in italics represent those existing in the thesaurus. In Fig. 11.12, it can be seen that many relations between these italicized words are found in the thesaurus. This suggests that the thesaurus can support the process of function synthesis in the FDP. In the trial, by using 'fan:vacuum cleaner–blow:suck up', as shown in Table 11.2, a lower level function 'fan blows dust', was obtained from an upper level function 'vacuum cleaner sucks up dust'. By using the superordinate-subordinate relations of the verbs 'rotate–move' and 'drive–move', along with the superordinate-subordinate relation of the nouns 'electric motor–motor', a new function 'motor moves impeller + blade' could be obtained by combining a lower level function of "hair clipper", such as 'motor drives blade', and a lower level function of "vacuum cleaner", such as 'electric motor rotates impeller'.

From the viewpoint of the aim of function synthesis, which is to generate a new function in order to answer the question: 'What should be created?', we need to extend this method to be able to obtain an *innovative* upper level function in Fig. 11.4, although a feasible new function structure is derived. In this case, a new upper level function which is beyond the categories of 'vacuum cleaner' and 'hair clipper' is expected to be generated. We believe that extending the method introduced in this chapter makes it possible to generate an *innovative* function beyond the categories of existing products.

References

1. Chiu I, Shu LH (2007) Using language as related stimuli for concept generation. AI EDAM 21:103–121. doi:10.1017/S0890060407070175
2. Gero JS (1990) Design prototypes: a knowledge representation schema for design. AI Mag 11:26–36

3. Hirtz J, Stone RB, McAdams DA, Szykman S, Wood KL (2002) A functional basis for engineering design: reconciling and evolving previous efforts. Res Eng Des 13:65–82. doi:10.1007/s00163-001-0008-3

4. Miles LD (1989) Techniques of value analysis and engineering, 3rd edn. McGraw-Hill, New York

5. Minami K, Taura T, Tsumaya A (2010) A study on modeling of reasoning process considering latent function. In: Proceedings of design and systems division conference of Japan Society of Mechanical Engineers. Tokyo, Japan, 27–29 Oct, pp 202–207 (in Japanese)

6. Pahl G, Beitz W (1995) Engineering design: systematic approach. Springer, Berlin

7. Taura T, Yoshikawa H (1992) A metric space for intelligent CAD. In: Brown DC, Waldron MB, Yoshikawa H (eds) Intelligent computer aided design. North-Holland, Amsterdam

8. Yamamoto E, Taura T, Ohashi S, Yamamoto M (2010) A method for function dividing in conceptual design by focusing on linguistic hierarchal relations. J Comput Inf Sci Eng 10:031004. doi:10.1115/1.3467008

9. Yoshikawa H (1981) General design theory and a CAD system. In: Sata E, Warman E (eds) Man-machine communication in CAD/CAM. North-Holland, Amsterdam

Chapter 12
Summary and Discussion

Abstract In this chapter, the previous chapters are summarized, followed by a discussion on steps to be taken in the future. First, the previous chapters are summarized into the following issues: consideration of *concept generation* (Chaps. 2–5), findings and inferences from the experiments and simulation (Chaps. 5–8), new research methods for *concept generation* (Chaps. 6 and 7), and the practice of *concept generation* (Chaps. 9–11). Next, a complete discussion is attempted on the basis of three themes: (1) what are the *inner criteria*? (2) what is an *ideal* in *concept generation*? and (3) how should the very early stage of design be investigated? Finally, some general perspectives towards the *future* of *concept generation* are suggested.

12.1 Summary

The previous chapters are summarized into the followings issues: consideration of *concept generation*, findings and inferences from the experiments and simulation, new research methods for *concept generation*, and practice of *concept generation*.

12.1.1 Consideration of Concept Generation

This book attempted to develop a systematized theory and methodology on the thinking process at the very early stage of design, in particular, in an interdisciplinary manner. Here, 'very' was used to exaggerate the beginning of design, which includes the time just prior to or the precise beginning of the so-called conceptual design, the very early stage of design was called the process of *concept generation*.

We used the term *concept* in this book in order to capture the very early stage of design in a general manner by focusing on the image generated by a designer.

T. Taura and Y. Nagai, *Concept Generation for Design Creativity*,
DOI: 10.1007/978-1-4471-4081-8_12, © Springer-Verlag London 2013

The scope of this book was fixed as follows. First, we limited the scope of the concept discussed in this book to concepts related to 'things', and other concepts such as 'time' or 'space' were treated as beyond the scope. Here, 'things' involve nonphysical objects, such as software, pictures and music, and physical objects: those which exist in a designer's mind as well as in the real world. Second, we assumed that a new concept is not generated from nothing. This declaration involves the following two assumptions: that the basis of the *concept generation* exists, and a new concept is generated by referring to some existing concepts which lie either in the real world or in a designer's mind. However, we do not deny that a new concept might be generated suddenly in a designer's mind with no foretokening, and we did not discuss this type of *concept generation* in this book owing to the difficulty in understanding these phenomena.

Within the above-mentioned scope, this book attempted to answer the questions: 'From where and how is a new concept generated?' and 'What enables a designer to promote the process to generate a new concept?'

The essence of *concept generation* was considered as follows.

In Chap. 2, previous research on *concept generation* was overviewed. The overview discerned that a critical process when producing a creative concept at the very early stage of design is shared among multiple design domains (engineering design, industrial design, etc.) as an unsystematized phenomenon, and deduced that the research on *concept generation* should be an interdisciplinary (beyond the existing academic disciplines and across multiple design domains) research theme. Bearing these perspectives in mind, some characteristics of the *concept generation* were discussed. First, *concept generation* was classified into two phases—the *problem–driven phase* and *inner sense–driven phase*—according to two factors: the basis of the *concept generation* and ability which enables the *concept generation* to proceed. Next, we defined *concept generation* and creativity: *concept generation* is the process of *composing* a desirable concept towards the *future*, and *design creativity* is the degree to which an *ideal* is conceptualized. In addition, the essential nature which enables a designer to promote the process of *concept generation* was discussed in Chap. 3 in terms of *design competence*. The *design competence* was categorized into three types: competence to inspire the motivation from inside a thought space, competence to abstract the concepts, and competence to control the *back-and-forth issue*.

On the basis of these fundamental considerations, a theory and methodology of *concept generation* were developed. In Chap. 4, a systematized theory of *concept generation* was developed. The theory classified *concept generation* into two types: *first-order concept generation*, which is based on the similarity recognition process, and *high-order concept generation*, which is based on the dissimilarity recognition process. The close investigation suggested that *first-order concept generation* is related to the *problem–driven phase* and *high-order concept generation* is related to the *inner sense–driven phase*.

In Chap. 5, *concept generation* was systematized into more specific methods of *concept synthesis*: *property mapping*, *concept blending*, and *concept integration in thematic relation*. It was shown that these methods correspond to *property mapping*, *hybrid*, and *relation linking* in the field of linguistic studies, respectively. This correspondence not only makes it possible to compare *concept synthesis* with linguistic interpretation, but also validates the classification of the three methods of *concept synthesis*, particularly the classification between *property mapping* and *concept blending*.

12.1.2 Findings and Inferences from the Experiments and Simulation

The findings and inferences from the experiments and simulation are summarized.

12.1.2.1 Cognitive Experiments

A total of four cognitive experiments were conducted.

The first experiment (Chap. 5) compared *concept synthesis* with the process of linguistic interpretation in order to determine the characteristics which feature in *concept generation*. This experiment revealed that the main factors in the *concept synthesis* are *concept blending* and *nonalignable difference*. This result gives certain relevance to the definition of *concept generation* (*composing* a desirable concept towards the *future*) that places emphasis on the *inner sense–driven phase*, which is related to *high-order concept generation*. Moreover, the result suggested that focusing on *nonalignable difference* is not a trait accumulated in subjects; rather, it occurs with regard to the design and interpretation processes.

The second experiment (Chap. 6) investigated the effect of the expansion of the thought space on the creativity of *design ideas*. The results found that there was a marginal significant correlation coefficient between the expansion of the thought spaces and the evaluated score of originality. These findings indicated that expansion of the thought space leads to a highly creative *design idea*. Furthermore, it was found that the *thematic relations* were used by the designer more frequently in the case of designing a *design idea* with a high evaluated score of originality than one with a low evaluated score of originality. From these results, it was inferred that expanding the thought space on the basis of the association process leads to a highly creative *design idea*; in particular, expanding the thought space by using *thematic relations* is an effective method of creating a highly creative *design idea*.

The third and fourth experiments (Chap. 8) determined the conditions for successful *concept synthesis* in *high-order concept generation*, particularly in *concept blending*. In the third experiment, the effect of the distance between the *base concepts* on the creativity of the designed outcomes was examined. The results revealed that the highest evaluated score of originality was obtained in the case of blending the *base concepts* with high distance. In the fourth experiment, the effect of the *associative concepts* of the *base concepts* on the creativity of the designed outcomes was examined. The results suggested that, in *concept blending*, the *base concepts* with a high number of *associative concepts* lead to higher originality in designed outcomes.

12.1.2.2 Computer Simulation

A computer simulation in order to capture the characteristics or patterns in the *concept synthesis* which may lead to a creative *design idea* was conducted (Chap. 7). This approach employed a research framework called *constructive simulation* which may be effective in investigating the generation of a concept—a process which is difficult to observe externally or internally. In *constructive simulation*, the difficult-to-observe mechanism which forms an observable real-life phenomenon is inferred from a different mechanism. When a phenomenon which is similar or analogous to the real-life phenomenon emerges from a different mechanism in the simulation, we believe this indicates that the two mechanisms share certain essential features. In the simulation, the virtual concept synthesis process was developed on a semantic network by tracing the relationships between its governing concepts.

From the simulation, the following characteristics or patterns of the actual concept synthesis process were inferred.

First, from the fact that we used the same semantic network to construct a virtual concept synthesis process for each designer, it was inferred that the creativity of the *design ideas* (*design idea features*) produced by individual designers does not originate from the differences in the structures of the concepts in their minds; rather, it originates from the manner in which their thinking proceeds. Accordingly, this suggests the possibility that there are common characteristics or patterns in the process of *concept synthesis* which effectively lead to a highly creative *design idea*.

Second, from the fact that the virtual concept synthesis process involves the notion of 'continuity' which is composed of the virtual concepts, it was inferred that the creative thinking process is originally continuous, but appears to be noncontinuous because it is partially hidden within the 'inexplicit' mind.

Third, from the fact that the paths were searched at the abstract level, it was inferred that abstraction plays an important role in the actual concept synthesis process.

Fourth, from the statistical results of our analysis of the virtual concept synthesis process, it was inferred that the thinking patterns in which both explicit and 'inexplicit' concepts are 'intricately intertwined' in a complex manner may lead to a creative *design idea*.

12.1.2.3 Cognitive Aspects of Successful Creative Concept Synthesis

On the basis of the findings of and inferences from the experiments and simulation described above, the cognitive aspects leading to successful creative *concept synthesis* can be characterized by the following notions: dissimilarity, association, and complexity.

'Dissimilarity' was pointed out to be essential for *high-order concept generation* in Chap. 4. In Chap. 5, the experiment revealed that *nonalignable difference* is an important factor in *concept synthesis*. Further, in Chap. 8, the highest originality score was obtained in the case of blending the *base concepts* with high distance. Here, it is reported, as described in Chap. 5, that *nonalignable difference* is related to dissimilarity, and the high distance between the *base concepts* also seems to be related to the notion of dissimilarity. This means that results obtained from the experiments in Chaps. 5 and 8 are consistent with and support the discussion in Chap. 4, and that 'dissimilarity' is an important aspect of creative *concept synthesis*.

'Association' is another important aspect of creative *concept synthesis*, from the two viewpoints of the associative thinking process and association of the concepts. Chapter 6 addressed the question of whether the thinking process 'relates' one concept to another is thought to be an essential factor in creative *concept synthesis*. In addition, the simulation in Chap. 7 involved the notion of 'continuity', since the virtual concept generation process is constructed as the set of links which are composed of the least distances between the words. These considerations regarding the experiment and simulation indicate that creative *concept synthesis* is a process of continuous association.

On the other hand, the effect of the *associative concepts* of the *base concepts* on the creativity of the designed outcomes was confirmed in Chap. 8. In addition, the relationship of the polysemy of design ideas with their evaluated originality score was discussed in Chap. 7; here, the process of association is assumed to give rise to the richness of the polysemy. The above considerations suggest that 'association' is another important aspect of creative *concept synthesis*.

Further, we suggest that 'complexity' is another important aspect of creative *concept synthesis*. Here, we would like to introduce the notion of 'expansion' into that of 'complexity'. The experiment in Chap. 6 showed the effect of the expansion of the thought space on the creativity of *design ideas*. In addition, the simulation conducted in Chap. 7 indicated the importance of complexity and expansion for the creative thinking process. These findings show the importance of 'complexity' for creative *concept synthesis*.

12.1.3 New Research Methods for Concept Generation

It is difficult to observe the thinking process of *concept generation* objectively, whether externally or internally, since a designer is assumed to be deeply engaged in his or her work during the creative process. Accordingly, developing a new research method and verifying its effectiveness should also be investigated. In this book, two methods were proposed. The first one was the *extended protocol analysis method*, which was introduced in Chap. 6. An experiment using this method was conducted in order to obtain the in-depth data pertaining to the expressed design activity. The second method was the computer simulation mentioned above. By applying these two new methods, reasonable results could be obtained; this suggests the effectiveness of the two new methods.

12.1.4 Practice of Concept Generation

We also sought the essence of *concept generation* by applying the methods of *concept synthesis* to more specific processes and by confirming their feasibility. In this book, the methods were applied to three applications: shape design, motion design, and function design. In Chap. 9, a method to synthesize the abstract shapes was developed. In Chap. 10, a method to synthesize motions in order to generate a creative and emotional motion was developed, and in Chap. 11, a method to synthesize functions in order to support the design of a new function was developed. Here, we would like to note that the three specific processes were synthesized on spaces which were different from the original spaces: the abstract shapes were synthesized on the space of the evaluation function; the motion, on the space of the wavelet coefficients; and the function, on the space of a thesaurus. In addition, it should be noted that these other spaces can be understood to be more abstract spaces than the original spaces: the space of the evaluation function is more abstract than that in which the specific shape is described, the wavelet coefficients' space is more abstract than that in which the specific motion is described, and the thesaurus contains words which are more abstract than those through which the function is described. These results of the practice suggest that specific concepts can be synthesized on a more abstract space, which shares the role of abstraction in *concept generation* with the discourse in Chaps. 3 and 7.

However, these applications did not contain the mechanism which corresponds to *inner sense*. The *inner sense* was expected to be reflected in the designed outcomes through the interaction between a computer and a designer.

12.2 Discussion

A complete discussion is attempted on the basis of three themes: (1) what are the *inner criteria*? (2) what is an *ideal* in *concept generation*? and (3) how should the very early stage of design be investigated?

12.2.1 What are the Inner Criteria?

We would like to focus on the results of the creativity evaluation in the four experiments. In all the four experiments, it was found that Kendall's coefficient of concordance showed significant concordance in originality, while three of the four experiments showed significant concordance in practicality. These results indicate that we have common *inner criteria* in our minds when evaluating creativity. This consideration explains the existence of *inner criteria*.

However, one question arises: Is the same kind of *inner criteria* active for the processes of *concept generation* and concept evaluation? The simulation for both the processes suggests a clue to this discussion as follows. The simulation in Chap. 7 revealed that an 'intricately intertwined process' plays an important role in the process of generating a creative concept. In addition, it was found that not the content of the process but the manner of the process affects the creativity of *design ideas*. This result indicates that the manner in which a concept is generated is identified as the *inner criteria*. On the other hand, the manner in which the generated concept is 'evaluated' is also found to be captured in the same way, according to the findings of our previous study [2]. In order to capture the nature of impressions which people receive from the products, we developed a method for constructing 'virtual impression networks' using a semantic network which is based on a very similar method as that introduced in Chap. 7. We were particularly interested in understanding the manner in which people form impressions, that is, 'like' or 'dislike'. Our results showed that it is possible to explain the difference between the feelings of 'like' and 'dislike' using several structural criteria of impression networks. The result revealed that the difference between 'like' and 'dislike' originates from the manner of the impression process rather than a specific image or shape; the difference between 'like' and 'dislike' is related to the difference in the structure of the impression network itself; the network of 'like' is more 'intricately intertwined' than that of 'dislike'. These findings and considerations indicate that the same kind of *inner criteria* is active for the processes of *concept generation* and concept evaluation.

However, it is a fact that a variety of *design ideas* are generated in the experiments. How can we explain the paradox that a variety of *design ideas* are generated from common *inner criteria*? We can explain the paradox as follows: the *variety* in *design ideas* originates from how specific concepts are used to implement the structure of the *concept generation* process, whereas the creativity of the generated *design ideas* is evaluated on the basis of the characteristics of the network structure without specific concepts.

In Chap. 2, we mentioned that a sense of *resonance* in the mind can be an *inner criterion* for an *ideal*. With regard to this, it is assumed that we do not *resonate* with specific concepts; instead, we *resonate* with the manner of a process. In other words, the state of *resonation* is assumed to be the inner state in which the thinking process is 'intricately intertwined'.

12.2.2 What is an Ideal in Concept Generation?

In this book, we used the term *ideal*. We would like to further discuss the meaning of *ideal* in *concept generation*.

In Chap. 2, we mentioned that a sense of *resonance* in the mind can be an *inner criterion* for an *ideal*. In addition, in Sect. 12.2.1, the state of *resonation* is assumed to be the inner state in which the thinking process is 'intricately intertwined'. Here, it should be noticed that this assumption is considered from the sole viewpoint of the inner state of people's minds. The notion of *ideal* in *concept generation* can also be discussed from other viewpoints.

First, let us refer to the related discourse on *ideal* in the context of design. As mentioned in Chap. 4, GDT defines *ideal knowledge* as one wherein all the elements of the entity set are known, and each element can be described by *abstract concepts* without ambiguity. The *ideal knowledge* is found to be a Hausdorff space which enables the design space to be a metric one, making it possible to effectively search for a solution, as mentioned in Chap. 3. This discussion suggests that the formation of *ideal knowledge* generates the potential to promote the design process. However, we should note that *ideal knowledge* cannot exist, since it is impossible for people to know all the entities and distinguish them without ambiguity. The fact that such a potential model cannot exist indicates the essence of the model on *concept generation*. The second case is the notion of 'particle' (mass point). The notion of 'particle' helped advance practical dynamics, which later became the basis for the development of engineering. However, we should note that the notion of 'particle' is no more than a notion. That is, an object which has mass but no volume cannot exist. From the above cases, it can be stated that the potential models may place a distance from the actual phenomena, while these models are extremely useful in explaining many actual phenomena.

On the basis of the above considerations, we can infer that the notion of *ideal* may be different from that of 'existable' or 'achievable'. Accordingly, the ideal design process or ideal design product should be that which need not necessarily exist.

We feel that an appropriate distance should be placed between an *ideal* and the real world. This declaration is consistent with our strategy in Chap. 10 to generate a creative and emotional motion on the basis of the hypothesis that a creative motion, which is beyond the normal bounds of human imagination, can produce emotional impressions which *resonate* with the human mind. Indeed, the result that this hypothesis was valid in the experiment sufficiently supports the above consideration on an *ideal* in *concept generation*.

12.2.3 How Should the Very Early Stage of Design be Investigated?

As mentioned in Chap. 3, *design competence* is thought to involve the following three types: competence to inspire the motivation from inside a thought space, competence to abstract the concepts, and competence to control the *back-and-forth issue*. *Design competence* is thought to be derived from the inside. This indicates the difficulty of observing *design competence* objectively in addition to that of externally or internally observing the creative process at the very early stage of design, which was pointed out in Chap. 7.

So far, many cognitive approaches investigating the inner state of a mind have been based on the stimulus–response framework. However, this framework is effective for investigating the inner mind only, when the inner mind itself does not change owing to the process which occurred and which was followed by the stimulus. This assumption is unsuitable for investigating *design competence*, the boundary and function of which are thought to change owing to receipt of the stimulus itself. Furthermore, from a fundamental point of view, 'intrinsic motivation' which is an essential element of *design competence* is assumed to not be activated by the stimulus given from outside. If so, how can we investigate *design competence*? One approach is to investigate the thinking process on the basis of another framework which is different from the stimulus–response one. Indeed, the four cognitive experiments introduced in this book did not employ the tasks on the basis of the stimulus–response tasks; instead, the subjects were asked to generate some new concepts without any specific stimulus. We did not examine the response to the stimulus, but examined the process or outcome of the *concept generation* itself; in addition, we developed new research methods in order to capture the difficult-to-observe process. In this book, two methods were proposed, as mentioned in Sect. 12.1.3.

Another research method is to identify the changing of the inner state in a positive manner, assuming that the inner state is always changing and its changing is an essential part of the inner state of a mind. Indeed, we conducted a case study on the basis of this consideration [1]. In this study, a designer merged his inner and outer perspectives himself. He could recognize the changes in his motifs when he examined the merged perspectives. We determined that his inner motif was observed by himself owing to this recognition of his changing motifs.

In Sect. 12.2.2, we mentioned that an appropriate distance should be placed between an *ideal* and the real world. Following this context, we think that there may be a way to develop a theory or methodology; that is, we pursue an *ideal* design process rather than induce the actual design processes conducted in the real world. However, this declaration does not intend to deny the actual design activity or the research method to investigate these activities; instead, we intend to point out the importance of developing another research method to pursue an *ideal*.

12.3 Towards the Future

Although *concept generation* is identified as the process of the very early stage of design in this book, this identification is based on a macroscopic view of design. If the design is captured from a microscopic view, at every stage, including the detailed design stage, the process of *concept generation* can be recognized. Therefore, in this section, some general perspectives towards the *future* of *concept generation* are suggested.

12.3.1 Beyond the Scope of this Book

Even though the scope of this book was limited, some considerations discussed here are expected to help extend the scope in the future. The second notion of abstraction in Chap. 3 is an example: 'abstract' in the term 'abstract painting'. This kind of abstraction is expected to be an essential element of the creative activity of human beings and could contribute to the extension of the discussion beyond the scope of this book. Similarly, the inside–outside and *back-and-forth issues* discussed in Chap. 3 are also expected to contribute to the extension of the scope of this book.

In the context of creativity, we are often interested in outstanding professionals. Although most of the subjects in the experiments introduced in this book were students, the essence of outstanding professionals is also assumed to be captured from the *design competence* discussed in Chap. 3.

12.3.2 Towards the Next Generation of Design

The word 'design' is being used increasingly in various societal contexts, for example, career design, company design, and community design. With a desire to cut across social stagnation and the hopelessness that overshadows us, the focus is on people who have the ability to compose and create.

However, thus far, 'design' has played the role of providing a method of acceleration for obtaining increased efficiency in our industrial society. For most designed products, the expectation is that they must be 'easy to use', 'convenient', 'cheap', 'consume low energy', or 'easy to understand', all of which involve the notion of 'efficiency'.

In contrast, there are other views on design, other than that on efficiency. These views are more deeply related to our spiritual dimension, namely, the 'better sense of well-being' or 'richness of the heart' of society. Here, we should never consider efficiency as solely a negative influence, but should respect the role it has to offer. However, if the times change such that we can be released from the sole belief in efficiency, other important meanings of design can arise, and it is these meanings that we wish to shed light on. If such a time does come, design can truly be discussed in greater depth, and we can view the society of the next generation in terms of a new perspective on design.

12.3.3 Approach to Understanding Human Beings

We considered the perspectives of *future* and *ideal*—that is, something that only human beings can conceive of should be focused on. It is thought that the *inner sense–driven phase* and *high-order concept generation* can be performed only by human beings, because human beings can recognize the notion of *future* by using language and manipulating higher *abstract concepts*. Although we confess that the boundary is a bit ambiguous, we suppose that other animals (such as monkeys) can probably perform a certain degree of the processes in the *problem–driven phase*. In fact, it is well-known that monkeys use tools during their food intake, a behaviour which is certainly one from the *problem–driven phase*. However, the process of *concept generation*, particularly with a focus on the *inner sense–driven phase,* is something that could never be carried out by monkeys. These are things that only a human can do: imagine a desirable concept, conceive concepts using abstract notions like the *future*, or *compose* a new concept with a high level of manipulation.

In other words, the considerations on *concept generation* are to help find an answer to the essential question: 'What is a human being?' An endeavour to carry out research on *concept generation for design creativity* is expected to help us develop an approach to comprehensively understand human beings.

References

1. Nagai Y, Taura T, Sano K (2010) Research methodology for internal observation of design thinking in the creative self-formation process. In: Taura T, Nagai Y (eds) Design creativity 2010. Springer, London
2. Taura T, Yamamoto E, Fasiha MYN, Nagai Y (2010) Virtual impression networks for capturing deep impressions. In: Gero JS (ed) Design computing and cognition '10. Springer, London

About the Authors

Toshiharu Taura is a vice-dean and professor in the Organization of Advanced Science and Technology (and professor in the Graduate School of Engineering) at Kobe University. He received his B.S., M.S., and Dr. Eng. degrees from the University of Tokyo, Japan, in 1977, 1979, and 1991, respectively. From 1979 until 1988, he worked as a mechanical engineer specialized in rolling mill at Nippon Steel Corporation. In 1992, he joined the University of Tokyo as an associate professor, and in 1998, he joined Kobe University as a professor in the Mechanical Engineering Department. He is currently working on several research themes that focus on the creative thought process of both engineering and industrial design, including interdisciplinary aspects of design science. He is a member of the Advisory Board, heads the Design Creativity SIG of the Design Society, and is a Fellow of the DRS (Design Research Society). He is also the recipient of the Best Paper Prize at the 2nd International Conference on Design Computing and Cognition (2006).

Yukari Nagai is a professor in the School of Knowledge Science at Japan Advanced Institute of Science and Technology. She graduated from Musashino Art University, majored in Design (M.A.), and conducted research abroad at the Creativity and Cognition Research Studios, Loughborough University (2002). She received Ph.D.s from Chiba University (2003) and the University of Technology, Sydney (2009). She was also awarded the Best Paper Prize at DESIGN2002 Conference and International Conference on Design Computing and Cognition (2006). She is a member of the Advisory Board and co-chair of SIG Design Creativity of the Design Society, and is a Fellow of the DRS. She has served as an editorial board member of the Journal of Engineering Design and contributed as guest editor to the special issue of the journal 'Artifact' (2008). Her research interests are design creativity, design knowledge, and creative cognition.

Index

A

Abduction, 30
Ability, 2, 13, 21
Abstract, 22, 101, 156
    abstract concept, 23, 27
    abstract form features, 115
    abstract shape, 115
    abstract the concepts, 21
Abstract concept, 23, 27
    define, 23
    description, 23
Abstract form features, 115
Abstract painting, 23
Abstract shape, 115
Abstract the concepts, 21
    abstract painting, 23
    art informel, 23
    geometrical picture, 23
    Impressionism, 23
    meta-reality, 23
    motif, 23
    organic picture, 23
    psychological reality, 23
Accented, 126
Achievable, 4, 158
Alignable difference, 33, 44
Analogical reasoning, 12
Analogy, 11
Analysing, 13
Appropriateness, 12
Art informel, 23
Associated concept, 51
Association, 73, 85, 100, 108, 155
Association process, 64

Associative concept, 46, 108
Associative Concept Dictionary, 46, 51,
    101, 109
Attribute, 13
Attribute Listing Method, 10
Attribute space, 25
Autonomy, 22
Autopoiesis, 22

B

Back-and-forth issue, 21, 37
    attribute space, 25
    beam's path, 25
    evaluation space, 25
    function space, 25
    General Design Theory (GDT), 25
    metric, 25
    preservation of the similarity, 25
    searching space, 25
    spatial issue, 25
Base concept, 32, 33
    base function, 134
    base motion, 126
    base shape, 115
Base design idea, 85
Base function, 134
Base motion, 126
Base shape, 115
Basis, 2, 13, 21
Beam's path, 25
Better sense of well-being, 160
Biologically inspired design, 125
Blended Shape Creation System, 116